Developments in Environmental Modelling, 12

Wetland Modelling

Series Editor: S.E. Jørgensen
                 Langkær Vænge 9,
                 Copenhagen
                 Denmark

Developments in Environmental Modelling, 12

Wetland Modelling

Edited by

William J. Mitsch
School of Natural Resources, The Ohio State University
Columbus Ohio, U.S.A.

Milan Straškraba
Biomathematical Laboratory
Biological Research Centre, Czechoslovak Academy of Sciences
Cĕské Budĕjovice, CZECHOSLOVAKIA

and

Sven E. Jørgensen
Department of Environmental Chemistry, Royal Danish School of Pharmacy
Copenhagen, DENMARK

Technical editor
Judy A. Kauffeld
School of Natural Resources, The Ohio State University
Columbus, Ohio, USA

ELSEVIER
Amsterdam — Oxford — New York — Tokyo   1988

ELSEVIER SCIENCE PUBLISHERS B.V.
Sara Burgerhartstraat 25
P.O. Box 211, 1000 AE Amsterdam, The Netherlands

Distributors for the United States and Canada:

ELSEVIER SCIENCE PUBLISHING COMPANY INC.
52, Vanderbilt Avenue
New York, NY 10017, U.S.A.

ISBN 0-444-42936-0 (Vol. 12)
ISBN 0-444-41948-9 (Series)

Printed in The Netherlands

CONTENTS

Preface
Contributors

PREFACE

The study of wetlands is a relatively new field and the modelling of these systems is still in its formative stages. Nevertheless, the editors felt compelled to assemble this volume as a first statement of the state of the art of modelling approaches for the quantitative study of wetlands. We took a global view of wetlands in this book, not only by including a wide geographic distribution of wetlands, but also by including papers on both freshwater and saltwater wetlands. Wetlands are defined as systems intermediate between aquatic and terrestrial ecosystems, and include ecosystems under a wide range of hydrologic and ecologic conditions. The wetland types discussed in our book reflect that heterogeneity, ranging from intermittently flooded wet meadows to permanently flooded shallow reservoirs and lakes. We also have included modelling examples from coastal salt marshes, shallow estuaries, mesotrophic bogs, reedswamps, forested swamps, and regional wetlands.

Ideas began for this book as early as the summer of 1984, when a session on wetland modelling was held as part of INTECOL's Second International Wetland Conference, held in Třeboň, Czechoslovakia. The idea for the book at first simmered slowly, then advanced to the front burner in late 1986. In the meantime, most of the contributions contained herein are the result of research carried out and models developed since 1984. Therefore we consider it a thoroughly up-to-date synthesis. The international flavor of the book is further demonstrated by the fact that this book was developed and written in the USA, USSR, Czechoslovakia, and Denmark, partially edited in the hotels and beaches of Italy, and published in The Netherlands.

We are particularly grateful to Judy Kauffeld, editor for the School of Natural Resources at The Ohio State University, for serving as the technical editor of this book. Her editorial skills and talents for Apple Macintosh desktop publishing made the transformation of rough manuscripts into polished book chapters seem easy. We are thankful to Ruthmarie Mitsch for her editorial assistance with many of the manuscripts as she caught our grammatical and spelling errors. The staff of the School of Natural Resources, under the direction of Jan Gorsuch, typed several of the manuscripts into the computer and assisted with the manuscript mailings. We are deeply indebted to the United States Information Agency and the Commission for Educational Exchange Between Denmark and the United States of America for the Fulbright grant support which was vital for the completion of this work. The work was also assisted by the Hydrobiological Laboratory in the Institute of Landscape Ecology and by the Biomathematical Laboratory of the Biological Research Center, both of the Czechoslovak Academy of Sciences. We are also appreciative of the confidence shown to us by Elsevier Science Publishers, particularly by Ken Plaxton, for agreeing to publish this work. Last but not least, we would like to thank the authors of the chapters for their fine contributions to this book.

William J. Mitsch
Columbus, Ohio, USA

Milan Straškraba
Céské Budějovice, Czechoslovakia

Sven E. Jørgensen
Copenhagen, Denmark

August 1987

CONTRIBUTORS

G. A. Alexandrov
Computation Centre
USSR Academy of Sciences
40 Vavilov Street
Moscow 117 333   USSR

Robert Costanza
Center for Wetland Resources
Louisiana State University
Baton Rouge, Louisiana 70803 USA

John W. Day, Jr.
Center for Wetland Resources
Louisiana State University
Baton Rouge, Louisiana 70803   USA

Michael J. Duever
Ecosystem Research Unit
National Audubon Society
Box 1877, Route 6
Naples, Florida 33999 USA

Carl C. Hoffman
Freshwater Biological Laboratory
Copenhagen University
Helsingørsgade 51
DK-3400 Hillerød   DENMARK

Charles S. Hopkinson
Marine Institute
University of Georgia
Sapelo Island, Georgia 31327   USA

Sven E. Jørgensen
Department of Environmental
   Chemistry
Royal Danish School of Pharmacy
2 Universitetsparken
DK-2100 Copenhagen DENMARK

Dimitri O. Logofet
Computation Centre
USSR Academy of Sciences
40 Vavilov Street
Moscow 117 333   USSR

Peter Mauersberger
Department of Hydrology
Institute of Geography and Geoecology
Academy of Sciences of the GDR
Muggelseedamm 260
Berlin GDR

William J. Mitsch
School of Natural Resources
The Ohio State University
2021 Coffey Road
Columbus, Ohio 43210   USA

Bernard C. Patten
Department of Zoology
University of Georgia
Athens, Georgia 30602 USA

Fred H. Sklar
Center for Wetland Resources
Louisiana State University
Baton Rouge, Louisiana 70803 USA

Milan Straškraba
present address:
Biomathematical Laboratory
Biological Research Center
Czechoslovak Academy of Sciences
Branisovska 31
370 05 Céské Budějovice,
CZECHOSLOVAKIA
previous address:
Hydrobiological Laboratory
Institute of Landscape Ecology
Czechoslovak Academy of Sciences
Céské Budějovice, CZECHOSLOVAKIA

Richard L. Wetzel
Virginia Institute of Marine Science
College of William and Mary
Gloucester Point, Virginia 23062 USA

Mary L. White
Center for Wetland Resources
Louisiana State University
Baton Rouge, Louisiana 70803 USA

1/ WETLAND MODELLING—AN INTRODUCTION AND OVERVIEW

Milan Straškraba
William J. Mitsch
Sven E. Jørgensen

Why Model Wetlands?

Wetlands and shallow bodies of water were once regarded as wastelands that were best diked and drained. Our literature contains many descriptions of wetlands as sinister and forbidding. They are now recognized as valuable ecosystems for the roles they play in the landscape. Wetlands act as hydrologic and chemical buffers between uplands and deeper water systems and as habitat for a wide diversity of desirable biota (Gopal et al., 1982; Mitsch and Gosselink, 1986). It is our firm belief that to manage these systems properly and to optimize their roles in the landscape, we must understand quantitatively how these systems work and what to expect when we disturb or change them. We must further understand how by changing one part of a wetland we affect the rest of the wetland and downstream systems as well. We believe that ecological modelling, which includes both conceptual and simulation modelling, offers a tool to describe, quantify and predict the behavior of these systems. This book is a result of that belief.

An International Effort

This book is also the result of an awareness by several international scientists that mathematical and conceptual modelling should be a major part of meetings being planned for scientists from throughout the world to deal with wetland ecology and management. This was first recognized by the organizers of a series of two international meetings on "Ecosystem Dynamics in Freshwater Wetlands and Shallow Water Bodies" held in the U.S.S.R. in 1981 and 1983 within the activities of SCOPE (Scientific Committee on Problems of

1

the Environment) and UNEP (United Nations Environmental Programme). Project leader Bernard C. Patten and his committee decided, properly we believe, that mathematical modelling should be a strong feature of that wetland activity. The work of the first meeting has been published as a technical report (Logofet and Luckyanov, 1982), while the findings of the second meeting will be published as a comprehensive SCOPE volume (Patten et al., in press).

Because of this success and interest in mathematical modelling as a tool for wetland researchers and managers, we saw the merit in exploring the subject of wetland modelling, which had at least been discussed at these two prior international forums and had been summarized in two reviews of freshwater modelling (Mitsch et al., 1982; Mitsch, 1983), on the program of the Second INTECOL Wetland Conference held in Třeboň, Czechoslovakia, during the summer of 1984. That meeting brought together 135 participants from 26 countries, all contributing to the understanding of the role of freshwater wetlands in our environment. The tradition of international meetings of scientists of different disciplines dealing with wetlands, created at the first International Wetland Conference held in Delhi, India, in September 1980 (Gopal et al., 1982), was enlarged both in the number of participants and the number of countries involved. Also, a broader view of wetland research aspects was covered in Trebon, particularly with the inclusion of mathematical modelling which had barely been represented at Delhi.

Wetland Model Development

Many countries have developed wetland conservation programs that reflect not only that these precious ecosystems should be conserved but also that there is a need for wetlands in the overall strategy of managing our environment. This represents what has been named ecotechnology or ecological engineering (Straškraba and Gnauck, 1985; Jørgensen and Mitsch, 1988) – the application of profound ecological knowledge to use (not abuse), modify, and support ecosystems for the benefit of both nature and humanity. This approach to wetlands encourages continued development of wetland modelling as one of the only tools available for managing quantitatively or semiquantitatively such complex systems as wetlands.

Wetland models were developed rather late compared to river, lake, or even terrestrial models. There are several reasons for this late development. Wetlands are defined as "lands transitional between terrestrial and aquatic ecosystems where the water table is usually at or near the surface" (Cowardin et al., 1979). This definition shows the high physical complexity of wetlands, covering both terrestrial and aquatic systems, as well as emergent properties specific to the transition zone between these two systems. This complexity is reflected in the difficulties encountered when attempting to understand, model and manage wetland ecosystems. Modelling of wetlands began in the mid-1970s (see Mitsch et al., 1982; Mitsch, 1983), when sufficient data on the functioning of wetland ecosystems became available. Since then, the number of wetland models published has steadily increased, due in part to our increasing interest and knowledge of wetlands.

The Contributions in this Book

Today, most types of wetlands have been modelled and wide experience has already been gained from these modelling studies. There are obviously still many gaps in our knowledge, but the editors believe that it is the appropriate time to attempt to distill the state of the art of modelling of these increasingly important ecosystems.

Various scientists were invited to write chapters on different aspects of wetland modelling or on a case study characteristic of today's wetland modelling. The contributions presented here are partially the result of a session on the state of the art of wetland modelling at the Třeboň meeting. While some of the papers in this volume (Chapters 3, 4, 5, and 11) are close to the material presented by the authors at that meeting, several others (Chapters 2, 6, 7, 8, 9 and 10) are original wetland modelling papers developed since the meeting. Under these somewhat complicated circumstances, the present volume does not fully reflect either the Třeboň modelling session or a comprehensive review of wetland modelling.

The volume does, however, include modelling approaches for a wide variety of wetlands, including northern bogs (Chapters 3 and 4), coastal marshes (Chapters 5 and 6), forested wetlands (Chapter 7), freshwater marshes and wet meadows (Chapter 8), and shallow reservoirs and lakes (Chapters 9 and 10). General principles about wetland hydrology as it relates to modelling are covered in Chapter 2, while a hierarchical modelling approach to a heterogeneous regional wetland system is described in Chapter 11. This volume also presents several different approaches to the modelling of wetlands, including models with an emphasis on wetland hydrology, models with an emphasis on biological productivity and processes, models used to design and summarize large scale research projects, and models used for wetland management. In the final chapter we attempt to summarize and conclude the results presented in the different chapters and to answer generally the questions: Where are we today in modelling wetlands? Which research needs do we face?

Wetland Hydrology

As a unifying theme, hydrology of wetlands is discussed in detail prior to the introduction of specific wetland models. The paper on wetland hydrology by **Duever (Chapter 2)** is an attempt to distinguish the major processes important to wetland hydrology. The corresponding models are of a rather general character, have not yet been formalized, and can be considered, at the most, conceptual models. They verbally describe generalities of wetland hydrology rather than specify mathematical process formulations. In this respect, it is also evident that generalizations are derived from a specific set of examples which the author has been able to recognize and study, and no generally accepted schemes have been achieved. We may find several more attempts in the literature to formalize wetland hydrology. Mitsch and Gosselink (1986) describe in some detail the generalizations that can be made about hydrologic processes in wetlands. Before that, Koryavov (1982) specified in mathematical terms the interrelations between river water level fluctuations and those of adjacent wetlands. It cannot be said that his derivation is very general, since he is also constrained by the specific examples he was able to study in detail.

Northern Peatland Models

Another extreme in hydrologic models of wetlands is represented in the paper by **Alexandrov (Chapter 3)** on a model of a raised bog. It represents an original contribution to hydrologic wetland modelling. It is devoted, however, to a particular wetland type—a raised bog, and to one particular problem—the explanation of its cupola-like shape. The author's original mathematical model is represented by two coupled spatially-distributed differential equations. The equations are based on hydrologic theory and explain from the same principles the creation of either a cupola-like shape or the ridge-pool complex shape of the bogs. The differentiation into these two widely differing shapes depends on initial conditions of the model solution, especially on bog size, and also on the amount of precipitation. On the other hand, simulations with bog sizes of 100 and 500 meters suggest that bog throughflow seems unimportant for the bog's differentiation. **Logofet and Alexandrov (Chapter 4)** present a dynamic biomass-nutrient model, also published elsewhere (Logofet and Alexandrov, 1984a,b), to study the interrelations between mosses and trees in raised bogs. The model consists of five aggregated compartments for trees, dwarf shrubs, grasses, mosses, and litter. The competition between mosses and trees was added to the model and it was shown that the principal conclusions about the evolution of the given transitional bog toward the forest phase were not changed. The simulations are run for a span of about 100 years.

Coastal Marsh Models

Two chapters present the state of the art of coastal wetland modelling. **Hopkinson, Wetzel and Day (Chapter 5)** reviewed six different simulation models of coastal marsh and estuarine systems. Included are: 1) the conceptual models of Mississippi River deltaic plain region, including both nature and urban centers, developed by Costanza and his colleagues; 2) an application of a storm water management model to evaluate effects of urbanization on nutrient runoff to a coastal swamp forest in Louisiana; 3) and 4) two long-term modelling programs dealing with salt marsh ecosystems on the southeastern coast of the U.S.; 5) a model of detritus, microbes, and consumers in the aquatic portion of a salt marsh; and 6) a hydrodynamic model of nutrient transport and transformations in a tidal freshwater marsh and river. **Costanza, Sklar, White and Day (Chapter 6)** discuss a spatial simulation model of coastal marsh succession consisting of 2,479 1 km^2 cells, each with eight state variables, simulated over the interval 1956-2050. Model goodness of fit is calculated at several resolutions and the effects of management options are evaluated.

Forested Wetland Models

Mitsch (Chapter 7) reviews the connections among hydrology, nutrients, and productivity for forested wetlands. Specific studies reviewed are primarily from cypress-tupelo and bottomland hardwood forested wetlands in the eastern half of the U.S. There are relatively few quantitative models that describe the interrelationships; but several qualitative models, independently developed but with the same "Shelford-type" parabolic limiting curve, have been suggested to explain the relationship between forest productivity and hydrology. A simple simulation model of nutrients, hydrology, and productivity of a forested wetland is developed to show the importance of flow-through and pulsing on these systems.

Freshwater Marsh/Wet Meadow Models

Jørgensen, Hoffmann, and Mitsch (Chapter 8) present a model of Danish freshwater emergent marshes and wet meadows. This is an important application of wetland modelling, because the ultimate goal is to use the wetlands as nutrients traps. Denmark has an intensive agriculture which, due to a heavy use of fertilizers, discharges significant quantities of nitrogen as nitrate and ammonium to the environment. One of the important methods in dealing with these non-point sources may be the use of wetlands for denitrification. This management strategy supports the maintenance of existing wetlands and even the restoration of already drained wetlands. A protection zone of wetlands along the banks of many streams is foreseen in future environmental planning.

Shallow Lake and Reservoir Models

Two papers, one by **Straškraba and Mauersberger (Chapter 9)** and one by **Jørgensen (Chapter 10)**, present a review of shallow lake and reservoir modelling, including many approaches applicable to shallower wetland systems. The first paper reviews modelling activities at academic research centers in Czechoslovakia and Eastern Germany. The models are presented briefly and their practical applications outlined, including a dynamic optimization model of reservoir eutrophication. Theoretical developments are based particularly on two approaches: cybernetic and thermodynamic. The cybernetic approach focuses on the mathematical representation of ecosystem adaptation, self-organization, and self-evolution by means of control theory. The thermodynamic approach uses notions of nonlinear thermodynamics for deriving theoretical shape for typical ecosystem functions. It is demonstrated that both approaches can be combined effectively, the thermodynamic focusing on the lower and the cybernetic on the higher ecosystem organization level, both being based on optimality notions. Jørgensen presents a eutrophication model of a shallow Danish lake in Chapter 10. Modelling of eutrophication is very well developed; the paper gives the state of the art of this type of model. The model presented has been used for predictive purposes and validated at a later time.

A Hierarchical Approach to Wetland Modelling

Patten (Chapter 11) reviews a systems approach used as a design for a Long Term Ecological Research program at the Okefenokee Swamp in Georgia. The overall study plan, which includes empirical studies and systems theory, is centered around a hierarchical model which, from largest system to smallest, includes the geographic region, the watershed, input drainage, the swamp-upland partition, major ecological communities (such as marshes, shrub swamps, *Taxodium* swamps, etc.), subhabitats, and subhabitat components. Preliminary carbon biogeochemisty models are also presented for a marsh, shrub swamp, and mixed cypress swamp. The distinction between systems and non-systems ecology is explained by an example of the climax concept. When understood as an abstraction in the sense of dynamic systems theory, this is a useful concept enabling ecologists to base ecosystems theory on principles known for other dynamic systems. But non-systems ecologists stress that real climax was not and will not be directly observed and therefore consider the concept of no use. They lose sight of the distinction between a concept (an idealization) and reality, and by sticking to reality and not considering concepts, they lose any possibility of theoretical advancement.

Conclusions

As ecological modellers and editors of this volume, we believe that wetland modelling will continue to develop as these resources continue to be recognized for their values. This book illustrates well that there is not yet "one way" to model wetlands. There are several important phenomena which make wetland modelling different from lake and stream modelling and worthy of its own set of paradigms and approaches and which, at the same time, make wetland modelling difficult to accomplish. We cannot, however, be satisfied by the present approaches and methods and we cannot trust our models too much. For example, Hopkinson, Wetzel, and Day (Chapter 5) show that the usefulness of a widely accepted analytical systems method for analyzing factors controlling ecosystem behavior, namely that of sensitivity analysis, is limited with complex non-linear systems. The results of sensitivity analysis have been shown to depend strongly on the initial abstraction or conceptualization of the ecosystem structure. For reasons such as this, practical as well as theoretical developments are necessary for the continued development of modelling (Straškraba and Gnauck, 1983, 1985; Jørgensen and Mitsch, 1983; Jørgensen, 1986).

A diversity of approaches is warranted at this stage in the development of wetland modelling and we believe that this book presents that diversity. Whether our interest is in the theoretical underpinnings of wetland ecology or in the wise management of these ecosystems, there is value in the systems approach and in the use of ecological modelling tools. We hope further that a volume such as this will help stimulate even more research into the development of the theoretical and management models for these valuable ecosystems.

References

Cowardin, L.M., V. Carter, F.C. Golet, and E.T. LaRoe. 1979. *Classification of wetlands and deep water habitats of the United States.* U.S. Fish & Wildlife Service Pub. FWS/OBS-79/31, Washington, D.C., 103 pages.

Gopal, B., R.E. Turner, R.G. Wetzel, and D.F. Whigham, editors. 1982. *Wetlands: Ecology and Management.* National Institute of Ecology and International Scientific Publications, Jaipur, India, 514 pages.

Jørgensen, S.E. 1986. *Fundamentals of Ecological Modelling.* Elsevier, Amsterdam, 389 pages.

Jørgensen, S.E. and W.J. Mitsch, editors. 1983. *Application of Ecological Modelling in Environmental Management, Part B.* Elsevier, Amsterdam, 438 pages.

Jørgensen, S.E. and W.J. Mitsch. 1988. *Ecological Engineering—An Introduction to Ecotechnology.* J. Wiley, New York (in preparation).

Koryavov, P.P. 1982. Mathematical modelling of wetland hydrology. Pages 297-310 in D.O. Logofet, and N.K. Luckyanov, editors. *Ecosystem Dynamics in Freshwater Wetlands and Shallow Water Bodies, Volume 2.* Center for International Projects, Moscow, USSR.

Logofet, D.O. and G.A. Alexandrov. 1984a. Modelling of matter cycle in a mesotrophic bog ecosystem. I. Linear analysis of carbon environs. *Ecol. Modelling* 21:247-258.

Logofet, D.O. and G.A. Alexandrov. 1984b. Modelling of matter cycle in a mesotrophic bog ecosystem. II. Dynamic model and ecological succession. *Ecol. Modelling* 21:259-276.

Logofet, D.O. and N.K. Luckyanov, eds. 1982. *Ecosystem Dynamics in Freshwater Wetlands and Shallow Water Bodies.* Proceedings of the international scientific workshop on July 12 to 26, 1981, held in Minsk, Pinsk, and Tskhaltoubo, USSR. Center for International Project GKNT, Moscow,

USSR. Vol. 1:312 pages; Vol. 2:324 pages.

Mitsch, W.J. 1983. Ecological models for management of freshwater wetlands. Pages 283-310 in S.E. Jørgensen and W.J. Mitsch, editors. *Application of Ecological Modelling in Environmental Management, Part B*. Elsevier, Amsterdam.

Mitsch, W.J., J.W. Day, Jr., J.R. Taylor, and C. Madden. 1982. Models of North American freshwater wetlands. *Int. J. Ecol. and Environ. Sci.* 8:109-140.

Mitsch, W.J. and J.G. Gosselink. 1986. *Wetlands*. Van Nostrand Reinhold, New York, 539 pages.

Patten, B.C. and others. in press. *Ecosystem Dynamics in Freshwater Wetlands and Shallow Water Bodies*. Proceedings of a Meeting in Talinn, USSR, August 1983.

Straškraba, M. and A.H. Gnauck. 1983. *Mathematische Modellierung Limnischer Okosysteme*. Fischer Verlag, Jena, 279 pages.

Straškraba, M. and A.H. Gnauck. 1985. *Freshwater Ecosystems: Modelling and Simulation*. Elsevier, Amsterdam, 309 pages.

2/ HYDROLOGIC PROCESSES FOR MODELS OF FRESHWATER WETLANDS

Michael J. Duever

The processes of a conceptual hydrologic model of a freshwater wetland are reviewed in detail. The unique hydrologic characteristic of wetlands is their narrow range of water table fluctuation above and below the ground surface which is dependent upon regional climatic and topographic characteristics and local site characteristics (microtopography, soil, vegetation) that create the necessary water depths and durations of inundation. The dominant processes controlling the distribution of water include atmospheric circulation, precipitation, evapotranspiration, and surface and groundwater flows. Surface and groundwater flows involve different portions of the same water mass. This water mass can be subdivided into surface water and three groundwater components: an unsaturated zone above the water table, an unconfined saturated zone below the water table, and a confined saturated zone below an impermeable aquiclude. One or another of these subdivisions may not exist at some sites or at different times at the same site. Most hydrologic differences among the various types of wetlands and shallow water bodies are of degree rather than of kind.

Introduction

At the most general level, the hydrologic cycle is the circulation of water from the earth's surface to the atmosphere and back again. While physical processes predominate, biological processes can also significantly influence the pathways involved and the rates at which water moves. Precipitation reaching the earth's surface can evaporate from land or water surfaces, run off to the oceans as surface water, seep into the earth, or return to the atmosphere via plant transpiration.

The economic importance of water has resulted in the development of a tremendous amount of information on the factors that control its distribution and movement. Numerous models have been utilized in the application of this fund of knowledge and have contributed to the successful resolution of many water management problems throughout the world. Major emphases of these applications have been on water supply for residential, agricultural, and industrial uses, and how to control water to minimize flood damage and/or open up otherwise "useless" land for development.

In the past, wetland management has normally involved stabilization of water levels at some elevation, either well below or above the ground surface. If fluctuations are permitted, they are controlled within certain limits in terms of water depth and timing. The variable degree and timing of water level fluctuation characteristic of natural wetlands is rarely compatible with the high degree of control desired when managing an area. As a result of these priorities, until recently there has been little interest in the hydrologic characteristics of most types of natural wetlands, and thus, little information has been generated on the topic. In addition, the physical hardship and/or expense involved in working in many of these systems and the subtlety of the processes involved has further hindered the development of an even rudimentary understanding of freshwater wetland hydrology.

Koryavov (1982) reviewed the status of water dynamics modelling applicable to wetlands and shallow bodies of water, stressing their importance in the decision making process. Important results of a successful mathematical modelling effort would be: (1) an ability to determine with a known degree of confidence the relative importance of different components of the hydrologic cycle and (2) an ability to forecast possible changes resulting from manipulation of these components. While significant progress has been made in modelling some major components of the hydrologic cycle and successful applications have been implemented, there is still much to be done before the real world water dynamics of wetlands can be said to have been successfully modelled. Koryavov (1982) felt that the dynamics of atmospheric moisture represented the least understood component, while surface and ground water dynamics have been much more extensively documented.

The conceptual hydrologic model shown in Figure 2-1 represents a not unfamiliar general statement about freshwater wetland systems that has relevance both to available field data and feasible management options for both natural systems and systems that have been altered by humans. The model illustrates major system components and their relationships to each other and the dominant external inflows and outflows. The system components are functionally, although not physically, isolated from one another. Surface and groundwater are different portions of the same water body, but the processes of evaporation, transpiration, and water flow affect each very differently. Also, wetlands are characterized by emergent vegetation, which is normally rooted in the groundwater zone and grows up through the surface water zone. The

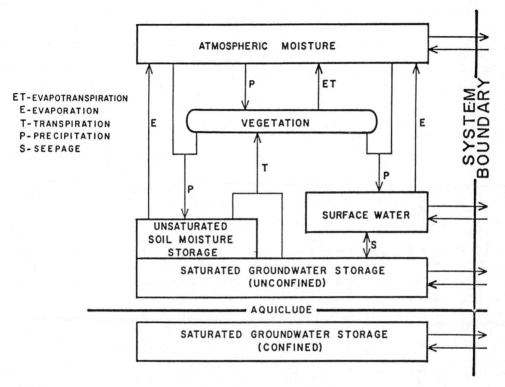

Figure 2-1. A conceptual model of wetland hydrology.

characteristics of different types of plant communities can variously influence evaporation, transpiration, and water flow rates of associated surface and groundwaters.

The objective of this paper is to describe the hydrologic characteristics and processes of freshwater wetlands according to this model. Particular emphasis is given to the general combinations of inflow-outflow characteristics that determine the existence of wetlands in a region and the more subtle intrasystem interactions that influence the distribution of wetlands within the general region. Since wetlands occur where the water table is close to the ground surface for much of the year during most years, comments on the intrasystem interactions will focus on the environmental conditions that produce these characteristic types of water level fluctuations.

Regional Hydrologic Characteristics of Wetlands

The regional occurrence of relatively stable water levels characteristic of wetlands can result from several alternative mechanisms. The simplest is a permanent surface or subsurface seep which flows into a shallow depression. This is most likely to occur where there is at least some topographic relief and where a sufficiently large watershed exists to provide a more or less constant supply of water regardless of the area's climate (Eisenlohr and others, 1972). In geographic areas with a cool moist climate, seepage wetlands can also develop on slopes as a result of low evapotranspiration rates and the water-holding capacity of peat substrates that develop in these situations (Lichtler and Walker, 1979). The other common wetland types exist along a precipitation-evapotranspiration gradient where stable water levels can result from (1) low evapotranspiration rates in low precipitation areas; (2) high precipitation in high evapotranspiration areas; or (3) some intermediate combination. Except at high altitudes or latitudes where precipitation is a more important influence, extensive wetlands are uncommon in areas with much topographic relief due to relatively rapid water losses associated with surface water runoff and evapotranspiration (Okruszko, 1967).

Thus, wetlands in temperate and tropical regions are found on relatively low-gradient topographic sites receiving high precipitation inputs. Low topographic relief reduces runoff losses, and high precipitation compensates for high evapotranspiration rates. Precipitation inputs can be either directly onto the wetland itself or through surface water flows from the surrounding watershed that overflow stream banks to supply floodplain wetlands or spill out onto shallow coastal areas to form deltas. Natural levees, accreted organic soils, dense vegetation, and other types of low physical barriers permit the rapid outflow of flood waters, but increasingly slow surface water flows out as water levels recede following high inflow periods.

The prime factor controlling the seasonal fluctuation in the vertical and horizontal extent of a wetland is its topography. This influences the area over which water can spread before depth increases to the point where runoff effectively removes any additional water inputs. Typically, vertical change in water level is relatively minor, generally less than 1 to 2 m, depending on the wetland type. A major exception is the floodplain wetland, such as those along the Amazon River, which can experience annual fluctuations of up to 20 m (Junk, 1983). While seasonal variation in the area inundated may be quite small, it is normally more variable from one site to the next than is seasonal variation in water depth. In some riverine situations, the width of the area inundated can vary seasonally by over 100 km. While all types of wetlands can exist over a wide range of sizes, shallow bodies of water are generally small and fluctuate by less than 1 km in horizontal extent. Just as wet and dry cycles of varying intensities occur over varying time periods, wetlands expand and contract or appear and disappear with these cycles. Wetlands receiving relatively large precipitation or runoff inputs are probably more buffered against these cycles.

Local Hydrologic Characteristics of Wetlands

While some major freshwater wetland systems exist as a result of relatively well-documented macroscale hydrologic conditions, differences in wetland community types or even the existence of numerous wetlands are dependent on much smaller scale hydrologic phenomena (Spence, 1964; Welcomme, 1979; Ingram, 1983; Duever, 1984). A particularly important aspect of these smaller scale phenomena is that they are readily alterable, either intentionally or unintentionally, by human activities and therefore must be a prime focus of wetland managers. Unfortunately, they are poorly documented for most wetland types.

The Flood Hydrograph

The flood hydrograph is a hydrologic model that represents the response of a certain portion of a watershed to water inflows and the dissipation of those inputs. While the shape and scale of a hydrograph can vary substantially depending upon the system involved, the basic configuration has general applicability to virtually all ecosystems where water occurs above ground for at least short periods during a normal annual cycle. In a typical hydrograph, initial water inflows are relatively small or non-existent, but at some point they begin to increase. This increase can be rapid and of short duration but relatively frequent during an annual cycle, or slow and of long duration. The time scales over which flood hydrographs operate vary from hours to months, while the spatial scales vary from less than a meter to 20 m vertically and from a few meters to 100 km horizontally. The predominant factor controlling these time and space scales is the size of the contributing watershed. Smaller watersheds tend to have frequent and very "spiky" hydrographs (Welcomme, 1979; Carter, 1979). Systems draining large watersheds tend to have a smoother annual hydrograph because of the spatial heterogeneity of their water sources and the more frequent presence of broad flat expanses which have a capacity to absorb and gradually release flood waters. Other factors control the timing of the flood hydrograph. In cooler climates, melting of winter snow accumulations results in increased water inputs and a spring flood hydrograph (Carter et al., 1979). In warmer climates, extended flood hydrographs are associated with an annual pattern of heavy wet season precipitation followed by many months of low rainfall (Welcomme, 1979; Duever et al., 1975).

Welcomme (1979) provided an interesting description of the long-term variability in water levels as synthesized from a composite of many flood hydrographs at one site. A particularly interesting aspect of his analysis was that the mean high-water stages and mean low-water stages predominated over either extreme or intermediate water level stages. We found a similar predominance of high and low water levels, as opposed to intermediate water levels, at Corkscrew Swamp in South Florida (Duever et al., 1978). The elevation of these mean high and low water stages, relative to local topography, can be expected to be crucial in determining an area's dominant kinds and distribution of biota.

Water Level Fluctuation

Water levels in depression seepage wetlands fluctuate as a function of the overall rise and fall of the surrounding water table. At these sites, the water table does not exhibit large annual fluctuations because water inputs are relatively constant. This constancy can be a function of a climate where water input and output relationships are relatively constant through the year, or a location where the wetland is associated with a relatively impermeable subsurface layer. Water percolating through the soil will encounter and move along the top of an impermeable stratum and emerge above ground where this stratum intersects the ground surface. It is in shallow depressions at or somewhere downstream of these sites that seepage wetlands can develop.

The larger and more common types of wetlands are dependent on a complex pattern of relatively microscale water movements. While very large total amounts of water may be involved because of the geographic extent of the wetland, water depths are shallow and water flows are very slow. In the Big Cypress Swamp in South Florida, Duever et al. (1986) found that, at the regional level, wetlands dominate much of South Florida because: (1) the flat topography reduces runoff to a minimum; (2) high rainfall during the warm part of the year more than compensates for high evapotranspiration rates; and (3) low evapotranspiration rates during the cool part of the year approximate precipitation inputs at this time. At the site-specific level, the character and extent of wetlands are determined by depth and duration of inundation, which are in turn influenced by each site's microtopography, soil type, and vegetative cover.

The normal cycle of water level fluctuation in wetland environments varies in its extremes, rates, and timing from one year to the next. Freiberger (1972) presents two hydrographs for a relatively undisturbed site which bracket the normal range of annual water level fluctuations at a single South Florida location (Figure 2-2). These hydrographs illustrate patterns (1) during a year (1957-58) with a fluctuation of less than 0.5 m because of heavy dry season precipitation, and (2) during a year (1970-71) with a fluctuation of over 1.5 m when a severe dry season drought occurred.

Monthly average water levels for 28 years of record (Figure 2-3) at the same site varied by less than 1 m (U.S. Geological Survey, 1979). Maximum average monthly water levels were even more stable, varying by only about 0.2 m, while minimum monthly water levels varied by up to 1.5 m. These data suggest that in a relatively high-precipitation area such as this, high water levels are possible at any time during the year, but droughts were very unlikely during the wet season. Reversals of this pattern of wet over dry season dominance probably occur in drier areas such as the prairie potholes region of the upper midwestern United States and southern Canada, where average annual precipitation is less than half that received in South Florida.

Water level fluctuations also vary as a function of the amount of aquatic or

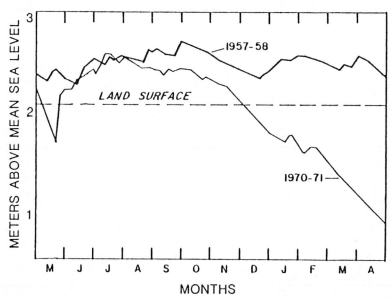

Figure 2-2. Hydrographs showing water levels at a site in the Big Cypress Swamp, Florida, during a year with a "wet" dry season (1957-58) and one with a "dry" dry season (1970-71) (Freiberger, 1972).

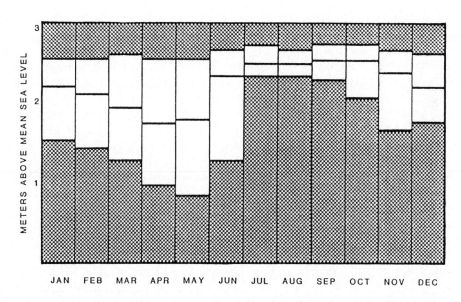

Figure 2-3. Water levels at a site in the Big Cypress Swamp based on 28 years of record (U. S. Geological Survey, 1979). The upper and lower edges of the clear area represent highest and lowest average monthly water levels, respectively, and the line represents average monthly water levels.

wetland habitat in the general area (Kulczynski, 1949). Fluctuations are minimal where surface water is always present, such as in large natural lakes, man-made impoundments and canals continually receiving water inputs, or sites along a coast. Water levels also tend to be relatively stable in wetland areas (Figure 2-4) because they have standing water much of the year and because of slow but steady runoff from upland areas for extended periods following rains. Upland sites exhibit more erratic patterns since they rarely have standing water and rainfall directly enters the soil. This produces sharp rises in the water table that quickly dissipate as the water drains off to lower areas. Canals can increase the frequency and size of these spikes on a hydrograph by generally lowering the water table below the ground surface for more of the time and by increasing the outflow rates following rains.

Even during high-water periods, groundwaters in low-lying coastal areas respond to the sinusoidal dynamics of tides (Welcomme, 1979). Such inland tidal movements are proportional to the amplitude and duration of the tide, distance inland, substrate storage coefficient, and transmissivity of the aquifer (Todd, 1959). Todd (1959) also demonstrated that there is a lag time prior to inland tidal response. Tidal effects on groundwater are usually insignificant in overall hydrologic patterns, but must be considered when evaluating data from wells located near a coast. Carter et al. (1973) found that a tidal range of 1 m produced a groundwater fluctuation of 0.9 cm at a point 3 km inland in South Florida.

An idealized hydrograph when water levels are more or less continuously declining through a severe dry season has three major components (Figure 2-5). The initial slow decline is represented by a gradual slope when the area is generally inundated. This normally occurs early in the dry season when surface runoff is the dominant process, but evapotranspiration is also significant. The middle portion of the hydrograph has a much steeper slope. The breaking point between the upper and middle portions of the hydrograph represents the period when the water table is dropping below ground over extensive areas in the vicinity of the well. When the water table is generally just below ground, it drops much more rapidly because evapotranspiration is still quite active, but the water only occupies about 30 percent or less of the soil volume. Thus, while a centimeter of water evaporated from standing water results in a centimeter decline in the water table, the same amount of water evapotranspired from the soil results in a decline of at least 3 cm. Since evapotranspiration is greatly reduced during cool winter months when plants are also relatively dormant, this slope is not as steep then as during the warm spring months when plants are actively growing. The final slow decline represents the period when the water table has dropped below the root zone and evapotranspiration generally ceases. At this point the slope is very gradual, being associated almost exclusively with groundwater runoff. This situation is observed occasionally during the later phases of severe droughts, but even then generally in higher sites only. The depth below ground at which the breaking point between the middle and lower portions of the curve occurs depends on the type of plant

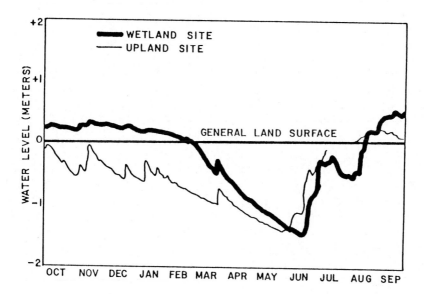

Figure 2-4. Daily hydrograph for two wells in the Big Cypress Swamp during water years 1972-73 (McCoy, 1974).

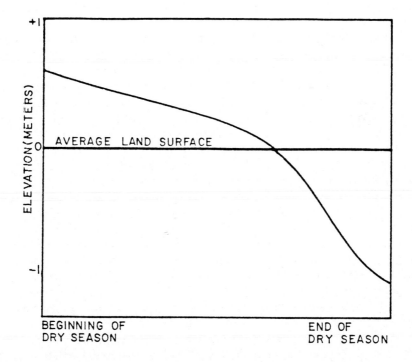

Figure 2-5. An idealized dry season hydrograph for a South Florida marsh.

community and substrate in the vicinity of the well. Herbaceous marsh communities in South Florida have few roots below 0.5 m while forested communities have root systems that penetrate more than 1.5 m (Duever et al., 1975). Romanov (1968) also interpreted an abrupt decline in evaporation from a raised bog as being due to recession of the capillary fringe below the rooting zone. The broad capillary fringe associated with fine textured soils, such as peat or muck, maintains locally higher water levels, which allows roots access to groundwater for much longer than in coarser soil types.

The relative importance of surface runoff, evapotranspiration, and groundwater runoff to the rate of water table decline can also be seen in terms of daily water level fluctuation patterns at a variety of water levels (Figure 2-6). The high rate of recovery at night when the water table is just below the ground surface is characteristic of a site with a nearby source of surface water. Sites without nearby surface water tend to show a relatively stable water level during the night when evapotranspiration rates are negligible.

Water Level Fluctuations in a South Florida Wetland

Duever et al. (1975) established four transects perpendicular to the main direction of flow through Corkscrew Swamp. Water levels were monitored weekly at five to nine wells along each transect for at least two years, and the results provided detailed information on some of the major factors affecting water levels. Figures 2-7 through 2-9 illustrate water table profiles along three of these transects for the 1974-1975 dry season. During the peak of the 1974 wet season, high rain fall rates had resulted in water table profiles sloping toward the swamp from the surrounding uplands. However, by late in the 1974 wet season, the water table along all three transects was depressed beneath the upland areas. This resulted from high evapotranspiration losses lowering the water table in the uplands and downstream flows maintaining high water levels in the swamp. As the dry season began during the fall, this same pattern became more pronounced. Most of the area was dry by February, when local differences in soil type began to dominate the water level profiles. The North Marsh transect (Figure 2-7) had fairly uniform sandy soils with extensive upland areas at each end, and its water table had achieved a relatively flat profile before it had declined far below ground. The Grapefruit Island transect (Figure 2-8) maintained a sloped profile as the water table declined below ground. This was probably a result of groundwater flows from the much more extensive upland areas on the up-slope end of the profile. However, extremely porous shell beds that underlay the down-slope end of the profile could also have accelerated flows in this direction. The Central Marsh transect (Figure 2-9) showed a much more complex dry season water table profile because of the presence of peat depths of up to 2 m in the lowland and large subsurface deposits of shell in the central portion of the transect. Relatively rapid percolation through the shell layers resulted in a pronounced localized depression of the water table throughout the dry season, while the peat deposits maintained a relatively high water table.

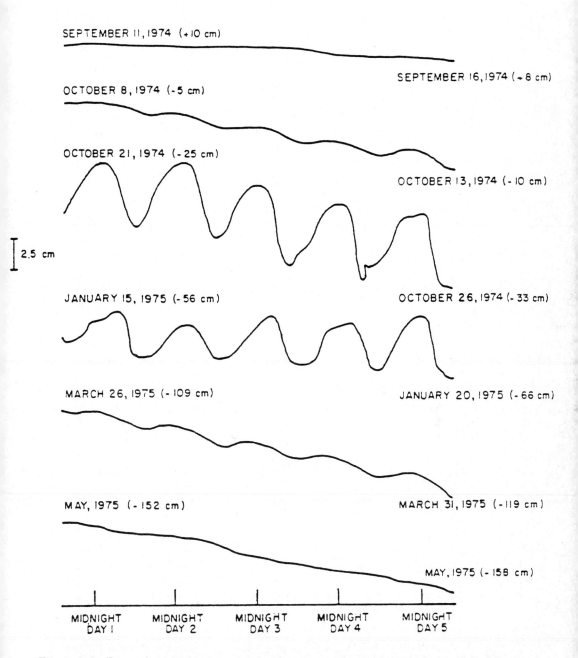

SEPTEMBER 11, 1974 (+10 cm)

SEPTEMBER 16, 1974 (+8 cm)

OCTOBER 8, 1974 (-5 cm)

OCTOBER 21, 1974 (-25 cm)

OCTOBER 13, 1974 (-10 cm)

2.5 cm

JANUARY 15, 1975 (-56 cm)

OCTOBER 26, 1974 (-33 cm)

MARCH 26, 1975 (-109 cm)

JANUARY 20, 1975 (-66 cm)

MAY, 1975 (-152 cm)

MARCH 31, 1975 (-119 cm)

MAY, 1975 (-158 cm)

MIDNIGHT DAY 1   MIDNIGHT DAY 2   MIDNIGHT DAY 3   MIDNIGHT DAY 4   MIDNIGHT DAY 5

Figure 2-6. Diurnal variation in water levels at an unforested upland site at Corkscrew Swamp, Florida (Duever et al., 1975). Numbers in parentheses indicate position of the water table in relation to the ground surface.

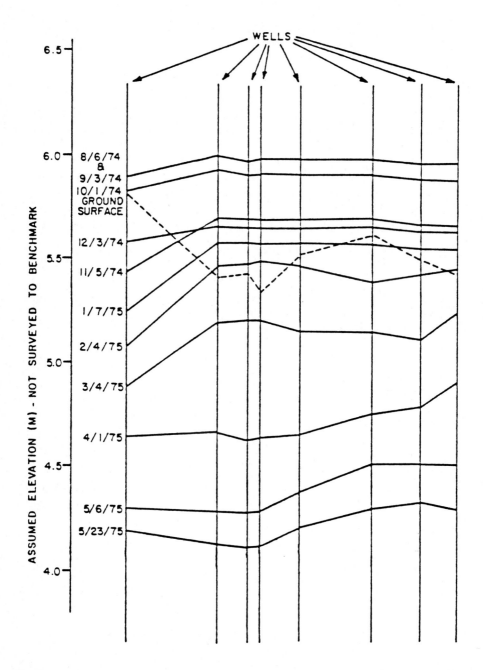

Figure 2-7. Water table profiles along the North Marsh transect at Corkscrew Swamp, Florida, during the 1974-75 dry season (Duever et al., 1986). The dashed line is the ground surface profile.

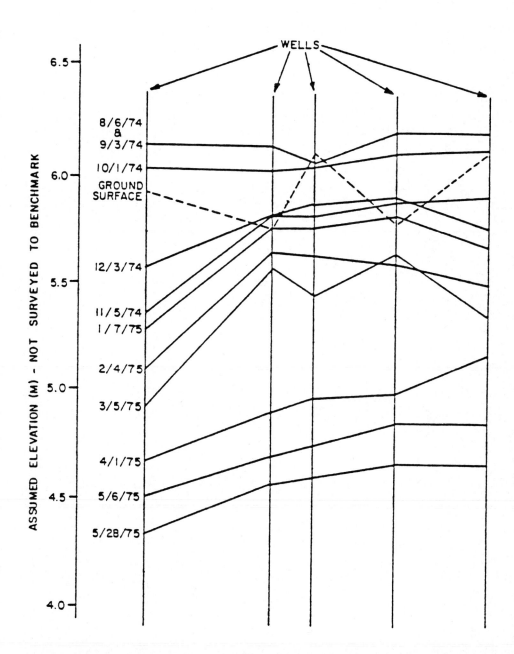

Figure 2-8. Water table profiles along the Grapefruit Island transect at Corkscrew Swamp, Florida, during the 1974-75 dry season (Duever et al., 1986). The dashed line is the ground surface profile.

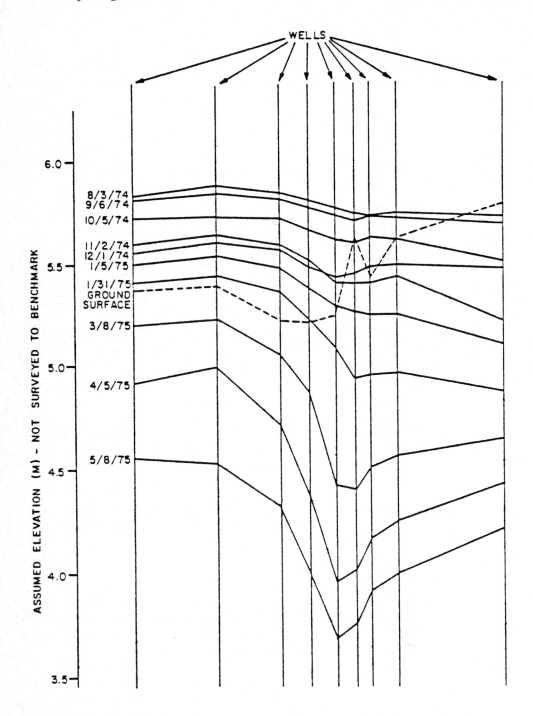

Figure 2-9. Water table profiles along the Central Marsh transect at Corkscrew Swamp, Florida, during the 1974-75 dry season (Duever et al., 1986). The dashed line is the ground surface profile.

Further indications of the affects of soil type on water levels when they fall below ground can be seen in the relatively large range of water levels at eight Corkscrew Swamp marsh sites during the dry season. At these times, marshes on mineral substrates exhibited the lowest water levels, while those on the deepest peat substrates had the highest (Figure 2-10). A similar pattern was found for nine cypress swamp sites underlain by mineral soil and peat substrates.

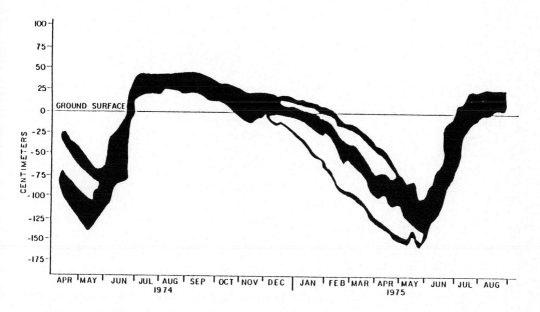

Figure 2-10. Range of water levels in eight marsh sites at Corkscrew Swamp, Florida (Duever et al., 1975).

Hydroperiod

Hydroperiod, or the annual period of inundation, is the summed expression of the temporal factors controlling wetland water levels. Although less well-documented than water levels, hydroperiod has been shown to be the dominant factor controlling both the existence and plant community composition of at least some major wetlands (Duever et al., 1986; Pesnell and Brown, 1977). It appears to exert direct control by eliminating species intolerant of extended inundation and indirect control by influencing frequency of severe fires.

Frequency of extended inundation is typically annual in wetlands and permanent in shallow bodies of water. Other environments with a greater than annual frequency of inundation generally tolerate only short-term, even if sometimes deeper and more frequent, flooding and support communities very different from those found on wetland sites. Annual duration of inundation must, by definition, be 12 months for permanently inundated shallow bodies of water and must be less than 12 months for the various wetland types. As latitude increases, the frequency and duration of frozen conditions also increases, so that wetlands can exist with correspondingly shorter hydroperiods. Thus, permafrost wetlands would typically have a hydroperiod of less than 3 months because of the short summers when water could even exist as a liquid. At the opposite extreme, tropical peatlands probably require hydroperiods in excess of 8 months to assure that annual rates of peat accumulation exceed or at least equal decomposition. Hydroperiods in mineral soil wetlands probably exceed 6 months in warm humid regions, but may be less is cooler climates or in what may marginally qualify as wetlands in more arid regions. Welcomme (pers. comm.) believes that floodplain wetlands are relatively insignificant along less than fifth order rivers and that the period of inundation along rivers of this size or larger typically lasts for approximately 2-6 months. Both latitude and precipitation patterns influence the duration of inundation on particular river floodplains.

Work at Corkscrew Swamp in South Florida has clearly shown that the distribution of undisturbed upland, marsh, swamp, and shallow aquatic habitats is largely a function of a site's hydroperiod (Duever et al., 1986). Some of these major habitats exhibit broad transitions in terms of hydroperiods, while the transitions between others and even the ranges of suitable hydroperiods are fairly narrow. Upland habitats at Corkscrew Swamp are typically inundated for less than 2 months annually and aquatic habitats for more than 10 months. The transition between upland and marsh habitats exists over a hydroperiod range of approximately 2 to 7 months with optimum marsh development at 7 to 9 months inundation and optimum swamp development at 9 to 10 months. Similar, even more subtle, variations in hydroperiod at Okefenokee Swamp have been related to the distribution of several major plant community types, although perturbations and successional processes are more involved in creating the full spectrum of major plant communities there (Duever, 1982).

Patterns of water level fluctuation determine a site's hydroperiod, and thus, the microhydrologic processes discussed above also determine the occurrence of hydroperiods suitable for the development of wetlands. The net effect of these processes is that wetlands quickly absorb water inputs to an area and only slowly release them through the year, while adjacent even slightly higher uplands lose water much more rapidly, resulting in either the absence of inundation or much shorter hydroperiods (Duever, 1982).

Long-Term Fluctuations

The above temporal and spatial patterns represent normal annual cycles in wetlands and shallow bodies of water. Relatively infrequent climatic events, e.g. floods or droughts, can produce more extreme patterns of water level fluctuation that result in major alterations in community structure and/or distribution. However, most wetlands are very tolerant of all but the most extreme floods and survive them with relatively minor and/or temporary overall effects. In some situations unusually high water levels may be necessary to recharge topographically isolated wetlands that would not exist without these occasional inputs of water. Drought can produce subtle changes in community composition by reducing numbers and distribution of species less tolerant of drought conditions. They can also have much more dramatic effects by providing conditions suitable for severe fires which can either selectively eliminate fire-sensitive species or completely consume the entire community along with accumulated organic soils. Alternately, droughts can provide the aerated soils and extended low water levels necessary for the germination and survival of some wetland plant species.

The areal extent, composition, and even existence of wetlands are all affected by long-term climatic cycles. These cycles differ from the sporadic occurrence of droughts and floods in that they represent gradual shifts of moisture conditions extending over decades, centuries, or longer. While climatic conditions return to a more normal range of variability following a major drought or flood event, long-term cycles produce gradual, but over time, major shifts in the normal year-to-year range of hydrologic conditions. Thus, as climatic cycles become wetter in areas suitable for their occurrence, wetlands will tend to dominate more of the landscape. There will also be a greater diversity of wetland communities because of a greater range of hydroperiods on different sites and the development of organic soils which create new edaphic sites. A major difficulty in managing wetlands is our inability to distinguish between, much less predict, shifts in hydrologic conditions that result from human activities and those resulting from sporadic natural events or long-term shifts in climatic conditions.

Wetland Hydrologic Processes

Solar Radiation

Solar radiation is the major force driving the overall hydrologic cycle. It provides the energy necessary for evaporation which results in atmospheric moisture that later falls as precipitation. Solar radiation also produces both local and global wind patterns by differential heating of the earth and its atmosphere. These wind patterns directly influence evaporation rates as well as the distribution of atmospheric moisture which is available for precipitation. Rates of snow and ice melt are determined primarily by solar energy inputs. Rapid melting can contribute significantly to severe floods, while slow melting can result in prolonged periods of high stream flows and wetland inundation.

Atmospheric Moisture

Atmospheric water represents a very small portion of the global water budget at any one point in time. However, large total annual quantities are involved because of its frequent turnover. Miller (1977) estimated that a molecule of water has an average residence time in the atmosphere of about 10-12 days, which while indicating fairly rapid turnover, is still a sufficient period to allow for long-distance transport. Most atmospheric water is in the form of a vapor (global mean content of 25 kg/m^2), and only small amounts are in a liquid (global mean content of 0.9 kg/m^2) or solid form (Miller 1977).

The world-wide distribution of water in the atmosphere is variable in both time and space. Amounts are normally greater during the summer and at lower latitudes because of the direct relationship between vapor pressure and temperature. Amounts also tend to be somewhat greater at times and places where surface water is more available, although this may be offset by relatively low water temperatures or wind patterns that bring relatively dry air masses into a region.

This component of wetland and shallow aquatic hydrologic systems interacts with other components of these systems primarily through precipitation (including fog and dew) and evapotranspiration processes. Atmospheric instabilities result in strong interactions with surrounding air masses in very complex short and long-term patterns. These air mass movements, temporal variability, and the existence of distinct air masses with varying moisture contents in the column of air overlaying a wetland site all make it very difficult to accurately measure actual quantities involved at a specific site. The most common technique for determining water content of an air column involves measuring relative humidity and temperature at a variety of points in the air column. Relative humidity is most commonly determined using psychrometers, and less frequently using dew-point or infra-red hydrometers (Oke, 1978).

Precipitation

Precipitation inputs to an ecosystem characteristically exhibit extreme spatial variability, even over relatively small areas during a single storm event. While recognizing this variability, virtually all studies synthesize an appropriate data set from the closest monitoring station(s) within or near the wetland and its watershed. Instruments used to measure precipitation and statistical methods involved in extrapolating field measurements over the area of interest, although imperfect, have been more or less standardized and information on them is available in a wide variety of references. In general, similar methods are utilized in all habitats.

Inputs to a ground or water surface beneath a vegetation canopy can be significantly different from those in an adjacent area without a canopy. Interception of precipitation by foliage and woody surfaces and re-evaporation of this water can significantly reduce the amount of water reaching the water table, particularly in parts of the world where light rains make up a major portion of the annual precipitation. Leyton et al. (1967) reported seasonal interception losses that approached 50 percent of precipitation in one study in England. Interception is a function of both patterns of rainfall events and the character of the vegetative surfaces involved (Ward, 1975). Interception, as a percentage of rainfall, is greatest when there is time for evaporation of intercepted rain before additional precipitation impacts the surface again. Thus, a pattern of many light rains would provide the greatest interception, and one of only a few heavy rains the least (Law, 1958).

Amounts of water intercepted are also related to the vegetative surface area to be wetted before water begins to flow off these surfaces. Thus, tall multi-strata forests can intercept more water than shorter or single-stratum communities. Also, evergreen coniferous forests intercept more water on an annual basis than do deciduous forests because the latter lose most of their surface area for a major portion of the year. Ward (1975) quoted interception values summarized by Rakhmanov of 25 to 35 percent of precipitation for conifers and 15 to 25 percent for deciduous forests. He also suggested that where total leaf area was similar in complete grass and shrub covers to that in a forest, interception amounts are similar. Ingram (1983) indicated that lichens could add to the interception capacity of a plant community.

Under foggy conditions, vegetation may strip water from the atmosphere, a certain portion of which can then drip to the ground. While this may represent a relatively small proportion of the total annual precipitation in some regions, in other areas it is reported to exceed precipitation (Kerfoot, 1968). Dew can also represent a small, but seasonally significant input. Lloyd (1961) found dew to be 7 to 21 percent of average monthly rainfall at a site in Idaho. Fog drip and dew are most significant on the edges of dense plant communities or when individuals are fairly well scattered. Both processes require some degree of air movement which is restricted in dense vegetation. Even more importantly, a moving air mass requires time for additional condensation to occur after its water droplets had been stripped by contact with vegetation.

Ward (1975) concluded his review of interception processes by indicating that there were discordant views about the significance of interception to water budgets: (1) interception and subsequent evaporation represents a net loss of water to the system; (2) interception and subsequent evaporation reduces transpiration losses and represents neither a gain or a loss to the system; and (3) interception and subsequent evaporation is more than compensated for by fog drip. It is likely that all three are true in the appropriate combinations of time and place.

Standard methods of determining canopy effects on precipitation normally rely on measurements of throughfall (amounts of precipitation passing through the canopy to the ground) and stemflow (amounts of water flowing to the ground on plant surfaces), and comparisons of these summed amounts to precipitation in a nearby area without a canopy. The difference is considered to be the amount of rain intercepted by vegetative surfaces and re-evaporated back to the atmosphere or, in the case of fog drip and dew, additional precipitation. When water budgets are constructed and interception losses are not specifically measured, they are normally considered to be part of total evapotranspiration when only area-wide precipitation and streamflow data are available.

In more northern climates, significant amounts of precipitation may fall during the winter as snow, but not be hydrologically available until spring when it melts. At these times, the large quantities of water released can contribute to rapid runoff when inputs exceed the capacity of the system to store or utilize these large amounts of water. As with precipitation, methods of measuring and estimating snowfall inputs have been fairly well standardized and are similar in different habitat types.

Evapotranspiration

Evapotranspiration is a major route by which water leaves natural ecosystems, and it frequently accounts for virtually all losses in wetlands because of their slow rates of flow and high surface area-to-depth ratios. Wetlands have higher evapotranspiration rates than many other habitats largely because they, by definition, store water near the ground surface where it is most susceptible to atmospheric losses. While some have considered this a "loss" of water, if wetlands were not available to initially store runoff, it would have been previously "lost" shortly after falling as precipitation.

Evapotranspiration is the combined process of evaporation from vegetation, land and water surfaces, and transpiration by plants. Evapotranspiration varies primarily as a function of microclimate (relative humidity, air and water temperature, wind velocity and duration), soil moisture content, and type and density of vegetation.

Relative humidity is inversely related to evapotranspiration since the higher the relative humidity, the closer the air is to being saturated with water, slowing the evapotranspiration rate. Higher temperatures increase the water-holding capacity of the air and the energy available for vaporization, both of which lead to more rapid evapotranspiration. Increased velocity and/or duration of wind results in higher evapotranspiration rates because air adjacent to the evaporating or transpiring surface is more rapidly replaced by less saturated air.

Soil moisture content is directly related to evapotranspiration, which obviously is at a maximum when standing water is present at or above the ground surface (Ingram, 1983). Soil type influences the distribution of moisture in the soil profile and hence evapotranspiration. Davis (1946) reported that Florida peat can contain approximately 350-935 percent water by weight. Because peat is so absorbent but water flows through it poorly, peat acts as a storage tank. Thus, water availability generally permits vegetation growing on peat to maintain a high evapotranspiration rate longer than similar plants growing on sand, marl, or rock.

Vegetation type significantly affects evapotranspiration rates, but in very complex ways associated with species, size, age, community structure, and geographic location (climatic conditions). In North Carolina, Swank and Douglass (1974) found that annual stream flow from a pine-covered watershed was 20 percent less than from a nearby watershed with a hardwood cover. The major reason for this was the more extensive leaf surface area of the pines, which resulted in greater evapotranspiration and interception losses. The higher soil moistures found in deciduous forests as compared to pine forests (Stoeckeler and Curtis, 1960) would further support this. Douglass (1967) and Rutter (1967) concluded that grasses use less water than trees because they are more shallowly rooted. Douglass (1967) also noted that evapotranspiration in humid regions varies directly with stand density and vegetative height. Baumgartner (1970) believed that increasing vegetation height increased air turbulence and thus evaporation. Alternately, Boelter and Verry (1977) found that peatlands covered by sedges and shrubs had higher evapotranspiration rates than forested peatlands. They attributed this to higher surface winds and a greater biomass of transpiring plants in the shrub and sedge community. Ward (1975) summarized data from Penman which indicated that forests of younger trees had higher total soil evaporation and transpiration rates, and that this combined rate began to decline after the trees were over 50 years old.

Baumgartner (1967) suggested that the differences in water consumption by various cover types are due to the differences in their albedo or reflectivity of short-wave radiation, which among vegetative types is lowest in coniferous forests. Belotserkovskaya et al. (1969) found year-to-year variations in evapotranspiration from the same bog plant communities that were inversely related to differences in albedo, which was higher during dry years. Fletcher and Lull (1963) compared evapotranspiration from forested sites with and without a litter layer, and bare soil with and without a litter layer. Bare soil covered by a litter layer had the lowest evapotranspiration rates. This was due to the elimination of transpiration, protection from wind and insolation by the litter, and the rapid decrease in evaporation as the water table declined through the soil profile.

Open bodies of water have high evaporation rates, which decrease with increasing amounts of emergent vegetation because of lower wind velocities and insolation inputs (Eisenlohr and others, 1972). In the upper midwestern United States prairie pothole region, during the peak of the growing season, evapotranspiration rates from a marsh exceeded open water evaporation. However, annual evapotranspiration was less than evaporation because of the short growing season. Sturges (1968) found that evapotranspiration from a Wyoming bog equalled or was somewhat greater than pan evaporation. In contrast to these relatively cool high latitude sites, Eisenlohr and others (1972) felt that total wetland evapotranspiration would probably increase more rapidly than evaporation from similar sized open bodies of water as one moved closer to the tropics due to the increasingly longer growing seasons. Croft and Monninger (1953) stated that on terrestrial sites transpiration is generally more important than evaporation because plant roots remove water from more of the soil profile than is available to evaporative processes. Rutter (1967) showed annual losses from a Scots pine plantation as a percentage of precipitation. They included 34 percent of precipitation to interception, 6 percent to soil evaporation and 46 percent to transpiration.

Evapotranspiration rates from whole ecosystems have been measured and calculated by a variety of techniques that have provided an amazing array of numbers. As indicated earlier, many factors influence evapotranspiration rates, and the seemingly infinite number of combinations of these factors makes it very difficult to either select an optimum methodology or detect meaningful patterns in the data that are available. Standard techniques have included: (1) evaporation pans containing various combinations of water, soil and vegetation; (2) theoretical estimates based on regional climatic data or site-specific microclimatic data; and (3) field measurements based on daily water level changes or precipitation minus downward seepage and/or runoff (Ward, 1975).

Water Flows

In general, knowledge of total annual water inputs provides limited information on a wetland's hydrology. Since the most important characteristic of a wetland is its extended, but not permanent, shallow inundation, the distribution of inputs through the annual cycle is a more important consideration. Analysis of data from Corkscrew Swamp in South Florida showed little correlation between annual precipitation and either wet or dry season water levels, but good correlations between dry season precipitation and water levels and between wet season precipitation and water levels (Duever et al., 1975). During high water level periods, large inflows can enter a wetland, but quickly dissipate as outflows or quickflows (Eisenlohr and others, 1972; Vecchioli et al., 1962). Several of these flood events occurring within a relatively short time span can substantially raise annual input totals, but represent merely transient events of little significance to the hydrology of most wetlands. However, these occasional peak flows are undoubtedly important to topographically isolated depression wetlands, which receive the majority of their inflows only during these events.

The water storage capacity of wetlands is intermediate between that of upland and aquatic habitats. Upland water storage is confined to soil interstices that represent approximately 30 to 60 percent of the soil volume (Lee, 1980), while 100 percent of the volume is available in aquatic habitats, and a variable combination of these volumes is available in wetlands. In each major habitat type, runoff rates drastically increase when water levels reach a point where the system's normal barriers to flow are overtopped. Simultaneously, the rate of water level rise falls off quickly as runoff rates approximate inputs. This leads to fairly constant year-to-year maximum water levels, particularly in wetland systems (Daniel, 1981). In upland habitats, this barrier is overtopped at the soil surface, while in aquatic habitats it occurs when normal outflow channels can no longer contain the flows. In wetlands, the barriers to flow are natural levees, relatively high land on an irregular topographic surface, accreted organic soils, and dense vegetation. The amounts of water confined and the base flows that occur after surface waters have receded below these barriers strongly influence wetland hydrology because of their control over the period of surface inundation.

A perched water table exists where a saturated soil layer is found above the regional water table and is separated from it by an unsaturated zone (Freeze and Cherry, 1979). This can occur where a relatively impermeable clay or organic soil stratum is near the ground surface and restricts the downward movement of water. Perched water tables can be found over a wide range of sizes. Some can continuously influence surface or near surface water levels over extensive areas, while others have only very local, temporary effects on water levels. However, regardless of their size, the hydrologic processes of importance to wetlands are basically the same as those described for sites lacking a perched water table.

A common misconception in wetland ecology is that wetlands typically occur on sites with perched water tables. While this may be the case in some situations, in the author's experience, wetland water levels have always coincided with the regional water table. Situations where I have encountered what appeared to be perched water tables have consistently turned out to represent unusual transient conditions. In several instances at the end of a long dry season, a marsh had over 15 cm of standing water while the water table was more than 50 cm below ground. This separation of water masses existed because the long drought had allowed the substrate to dry out sufficiently so that it took time for the near-surface soils to be rewetted before the surface water, which was produced by high intensity storms, could rapidly percolate to the water table. Another instance involved a forested wetland site with an organic substrate that was essentially dry at approximately 30 cm below the ground surface but which was overlain by standing water. My only explanation for this condition was that daytime evapotranspiration rates were faster than the movement of water down from the ground surface. At another site, 3 out of 31 monitoring wells consistently showed higher surface water levels compared to water levels monitored in 3 to 6 meter deep wells at the beginning of the dry season. This divergence continued to increase until water levels dropped below the ground surface. The 3 wells that exhibited this pattern were located over a subsurface shell bed that apparently functioned to accelerate horizontal groundwater outflows more rapidly than water could percolate down from the ground surface. Eisenlohr and others (1972) attributed the misconception about perched water tables in North Dakota prairie potholes to the relatively impermeable soils that were slow to equilibrate with changes in surface water levels in nearby potholes. The water table is normally considered to represent a muted version of the land surface, such that it is relatively far below ground in locally elevated sites, and above ground in locally depressed sites. These locally depressed areas, where the water table is above ground, are where wetlands are most likely to develop if water depths are not excessive. Lakes will occur if water depths are too great to allow the formation of wetlands.

In concert with the concept that wetland water levels reflect local water tables, there is little basis for looking on wetlands as significant groundwater recharge areas. If anything, they are more often associated with groundwater discharge (Holzer, 1973; Vecchioli et al., 1962). The relatively impermeable soils that characteristically underlie wetlands also argue against any significant groundwater recharge (Carter, 1979). Eisenlohr and others (1972) related most seepage losses from prairie potholes to evapotranspiration losses in the surrounding uplands and only a relatively small portion to downward seepage.

Groundwater Flows

Groundwater can be subdivided into three components. The uppermost is not always saturated with water, but the two lower components are normally saturated. The upper two are unconfined and more or less freely interact with water at the ground surface, while the lowest is confined and may be represented by one or more water masses separated from free interaction with the ground surface by impermeable geologic strata. Typically, the water table represents the boundary between the saturated and the unsaturated zones, although in fine-textured soils such as peat or clay the saturated zone can extend as a capillary fringe up to 40 cm above the water table (Ingram, 1983). An unsaturated substrate is a normal but transitory condition in wetlands, while in aquatic habitats the soil surface layer is normally saturated, and in terrestrial habitats it is normally unsaturated. The existence of an unsaturated soil layer significantly alters hydrologic flow patterns. Under these conditions, water is still present in the soil, but it is less available for involvement in hydrologic processes such as evapotranspiration or gravity flow. Indeed, when water potentials become low enough, adsorptive bonding to particle surfaces leaves any remaining soil water functionally dissociated from the hydrologic cycle.

The main soil parameters of significance to groundwater hydrologic processes are porosity and permeability. A highly porous soil will hold or store large amounts of water, while a highly permeable soil will allow large amounts of water to flow through it because of high connectivity among the pore spaces. The size of the pores, however, can also be important. A large number of small pores may hold large quantities of water, but the water may move slowly even if the pores are highly interconnected. The high surface tensions associated with numerous small pore spaces in peat or clay soils are the basis for their pronounced capillary fringe which is a major factor in the development and persistence of some wetlands. When soil profiles are not completely saturated, the storage component of a system's water budget is less than full. This condition also exists for the surface water mass in wetlands when normally inundated ground surface depressions are incompletely filled. These temporal variations in storage are relatively simple to account for because the volumes, flow directions, and rates can be measured by fairly straight-forward techniques. However, estimation of temporal variations in subsurface unsaturated soil storage is greatly complicated by a number of additional processes. These include infiltration of precipitation into soils, percolation through the soil, height of the capillary fringe above the water table, and availability of water for evaporation from the ground surface and for transpiration by plants. They are further complicated by horizontal and vertical variations in soil types.

There are two hydrologically important thresholds associated with the amount of water contained in unsaturated soils (Brady, 1974). The higher threshold is illustrated by the concept of field capacity which describes the amount of water a soil is capable of retaining against gravitational forces. Water added to an unsaturated soil at field capacity will directly contribute to raising the water table by downward percolation, but if added to a soil below field capacity it will be retained in the unsaturated zone until field capacity is attained. Light rains, even when frequent, may thus merely increase surface soil moisture temporarily until evapotranspiration processes return it to the atmosphere, but heavy rains will frequently exceed the field capacity threshold and provide at least some recharge to the water table. Another important threshold is the wilting coefficient which is a measure of the amount of water that still remains in the root zone of a soil profile, but is unavailable for evapotranspiration.

Evapotranspiration rates decline as the water table recedes below the ground surface and continues to decline as the water table recedes through the root zone. It effectively ceases when the portion of the soil profile within the root zone reaches the wilting coefficient. It is highly unlikely that permanent wetlands would experience drought of this severity, although this may be a frequent or even normal condition for some ephemeral types of wetlands or shallow bodies of water. Since both of these thresholds are a function of soil porosity and permeability at any one site, their expression is again a complex function of the horizontal and vertical distribution of soil types.

The lowermost groundwater unit normally involves one or more aquifers separated from the water table by impervious strata that severely limit the vertical movement of water. In reality, the confining beds rarely eliminate all vertical movement, and typically there is some exchange with unconfined groundwater associated with the water table. While confined aquifers can potentially contribute water directly to surface or groundwaters of wetlands and shallow bodies of water, there are probably relatively few situations where they add significant amounts compared to inflows from precipitation or unconfined water table aquifers.

Groundwater movements in wetlands are a result of (1) inputs from precipitation (throughfall and stemflow on vegetated sites) when surface water is absent, (2) seepage losses to surface water during high rainfall periods and gains from surface water during low rainfall periods, (3) transpiration losses, (4) evaporation losses when surface water is absent, and (5) groundwater inflows and outflows. Precipitation inputs have already been discussed and are merely partitioned into surface or groundwater compartments, based on the position of the water table. Vertical seepage to or from groundwater is extremely difficult to measure directly (Ingram, 1983), largely because it is a diffuse flow across an invisible boundary. In many studies, these flows are assumed to be either insignificant or in balance (on an annual basis), so that their overall effect on a water budget can then be ignored. The presence and general contribution of seepage to total inflows into surface water can be directly estimated by comparing water quality of surface inflows and outflows with similar water quality data from "upstream" groundwaters.

There is no simple way to directly determine seepage losses to groundwater, although isotope or chemical tracers could be useful. The difficulty of separating evaporation from transpiration losses normally dictates that they are determined as a single value, which is a function of a number of factors discussed earlier. Groundwater flows into and out of a wetland are typically most significant along its periphery (Millar, 1971; Eisenlohr and others, 1972) and probably decrease with distance from the upland edge, as was found for groundwater movements into a Canadian lake (Frape and Patterson, 1981). This is associated with the frequent occurrence of relatively impermeable peat and clay subsurface strata in wetland environments. Vecchioli et al. (1962) reported artesian flows through these bottom strata in New Jersey's Great Swamp, although the quantities involved were insignificant. Peripheral exchanges can be either into a wetland during high rainfall periods or out of it during extended low rainfall periods (Bay, 1967; Duever et al., 1975; Duever, 1982). They can be quantitatively estimated from information on water table slopes and soil transmissivity.

Water stored in saturated groundwater aquifers can be estimated from soil storage coefficients and definition of the area involved. Two common approaches to measuring water quantities in unsaturated strata utilize soil moisture tensiometer or neutron-scattering techniques, although there are major problems with their application in organic soils. These determinations are invariably complicated by the diversity of soil types present in wetlands. Any of the above quantities can be determined by difference if the others are known.

Surface Water Flows

In general, surface water movements in wetlands are a result of processes that are similar to or complement those described for groundwater: (1) inputs from precipitation (throughfall and stemflow on vegetated sites) when there is surface water, (2) seepage losses to groundwater during low rainfall periods and gains from groundwater during high rainfall periods, (3) transpiration losses when surface water is present, (4) evaporation losses when surface water is present, and (5) surface water inflows and outflows. Transpiration from surface water is associated only with vegetation that either floats on the water surface or, despite being rooted in the substrate, still sends roots out into the water column. Estimation of surface water storage requires a knowledge of site microtopography and the range of water level fluctuation. Since wetland water depths are frequently less than 1 meter, even during high water periods, microtopographic data with 0.1-0.2 meter contours are probably necessary for reasonably detailed estimates of surface water quantities. The most efficient means of deriving these data over extensive areas is by correlating major vegetation types with water depths and then using aerial photography to map and quantify the areas occupied by the different vegetation types. This approach is probably more accurate in the determination of water depths in relatively undisturbed areas than are extensive topographic surveys.

Since wetlands normally contain few if any streams, rates of surface water flow are essentially zero when water levels are low and thus contained by topographic irregularities. In South Florida marshes, Leach et al. (1972) measured average flow rates of 3.0 mm/sec during 1960 and a total range of approximately 0 to 5.4 mm/sec. These rates were determined over a range of discharges representative of those that had been recorded over the previous 20 years. Total distances moved during a dry year would be about 10.5 km and during a wet year about 65 to 80 km. Parker (1974) stated that these slow flows were due to a combination of the gentle gradient and the dense vegetation through which the water must move.

Conclusions

While all aspects of the conceptual hydrology model could be improved by additional data, certain aspects are supported by both fewer data and a lack of consensus about the relevance of what data is available. With the exception of some cool-latitude mire systems, most of the available information on vegetation or vegetated sites was derived from studies of terrestrial sites, while most of the information on surface water processes was derived from sites lacking emergent vegetation. Thus, measurements of most parameters in a number of very different wetland sites would be valuable in verifying the applicability of these data. Precipitation is probably the best documented process and we are at least generally aware of the limitations of the methodologies involved. Groundwater flows are generally considered to be relatively unimportant to wetland processes. The amount of work involved in a detailed accounting of groundwater flow processes probably could be better spent on other processes that account for much larger flows. A lack of data on the dominant ecosystem level processes that control patterns of water level fluctuations, particularly shallow surface and near-surface water movements and evapotranspiration rates, represents the most significant gap in our knowledge of natural wetland systems.

Finally, while a general well-documented model is of use in orienting a manager to his particular system and its important parameters, it is rarely adequate for addressing management problems without a reasonable amount of site-specific data to verify where the model needs to be modified. These modifications have characteristically been necessary because most wetland areas have been altered by man to some extent. These alterations must be accounted for before the model can be truly applicable.

Acknowledgment
This work was supported by a subcontract from the University of Georgia Institute of Ecology under NSF Grant Award DEB 76-12292. This is University of Georgia, Okefenokee Ecosystem Investigations Publication No. 68.

References

Baumgartner, A. 1967. Energetic bases for differential vaporization from forest and agricultural lands. Pages 381-389 in W.E. Sopper and H.W. Lull, editors. *Forest Hydrology.* Pergamon Press, Oxford.

Baumgartner, A. 1970. Water and energy-balances of different vegetation covers. Pages 56-65 in *World Water Balance*, International Association of Scientific Hydrology. Gentbrugge.

Bay, R.R. 1967. Factors influencing soil-moisture relationships in undrained forested bogs. Pages 335-343 in W.E. Sopper and H.W. Lull, editors. *Forest Hydrology.* Pergamon Press, Oxford.

Belotserkovskaya, O. A., I.F. Largin, and V. V. Romanov. 1969. Investigation of surface and interval evaporation on high-moor bogs. *Soviet Hydrology* 1969: 540-554.

Boelter, D.H. and E.S. Verry. 1977. *Peatland and water in the northern lake states.* U.S. Forest Service Gen. Tech. Rep. N.C. 31. North Central Forest Experiment Station, U.S. Forest Service, St. Paul, Minnesota, 22 pages.

Brady, N.C. 1974. *The Nature and Properties of Soils, 8th ed.* McMillan Publishing Company, New York. 639 pages.

Carter, M.R., L.A. Burns, T.R. Cavinder, K.R. Dugger, P.L. Fore, D.B. Hicks, H.L. Revells, and T.W. Schmidt. 1973. *Ecosystems analysis of the Big Cypress Swamp and estuaries.* South Florida Ecological Study, Surveillance and Analysis Division, U.S. Environmental Protection Agency, Region IV, Atlanta, Georgia, 375 pages.

Carter, V. 1979. Hydrologic and hydraulic values. Pages 52-79 in J. Clark and J. Clark, editors. *Scientific Report, The National Symposium on Wetlands.* National Wetlands Technical Council, Washington, DC.

Carter, V., M.S. Bedinger, R.P. Novitzki, and W.O. Wilen. 1979. Water resources and wetlands. Pages 344-376 in P.E. Greeson, J.R. Clark, and J.E. Clark, editors. *Wetland Functions and Values: The State of our Understanding.* American Water Resources Association, Minneapolis, Minnesota.

Croft, A.R. and L.V. Monninger. 1953. Evapotranspiration and other water losses on some aspen forest types in relation to water available for stream flow. *Transactions of the American Geophysical Union* 34:563-574.

Daniel, C.C., III. 1981. Hydrology, geology and soils of Pocosins: a comparison of natural and altered systems. Pages 69-108 in C.J. Richardson, editor. *Pocosin Wetlands. Proceedings of Pocosins: A Conference on Alternative Uses of the Coastal Plain Freshwater Wetlands of North Carolina.* Hutchison Ross Publ. Co., Stroudsburg, Pennsylvania.

Davis, J.H., Jr. 1946. *The peat deposits of Florida their occurrence, development, and uses.* Fla. Geol. Surv. Geol. Bull. 30, Tallahassee, Florida, 247 pages.

Douglass, J.E. 1967. Effects of species and arrangement of forests on evapotranspiration. Pages 451-461 in W.E. Sopper and H.W. Lull, editors. *Forest Hydrology.* Pergamon Press, Oxford.

Duever, M.J. 1982. Hydrology–plant community relationships in the Okefenokee Swamp. *Florida Scientist* 45:171-176.

Duever, M. J. 1984. Environmental factors controlling plant communities of the Big Cypress Swamp. Pages 127-137 in P. J. Gleason, editor. *Environment of South Florida: Present and Past. II.* Miami Geological Society, Miami, Florida.

Duever, J.J., J.E. Carlson, and L.A. Riopelle. 1975. Ecosystem analysis at Corkscrew Swamp. Pages 627-725 in H.T. Odum, K.C. Ewel, J.W. Ordway, and M.K. Johnston, editors. *Cypress wetlands for water management, recycling, and conservation.* Second annual report to National Science Foundation and Rockefeller Foundation. Center for Wetlands, University of Florida, Gainesville.

Duever, J.J., J.E. Carlson, L.A. Riopelle, and L. C. Duever. 1978. Corkscrew Swamp. Pages 534-570 in H.T. Odum, K.C. Ewel, editors. *Cypress wetlands for water management, recycling, and conservation.* Fourth annual report to National Science Foundation and Rockefeller Foundation. Center for Wetlands, University of Florida, Gainesville.

Duever, M.J., J.E. Carlson, J.F. Meeder, L.C. Duever, L.H. Gunderson, L.A. Riopelle, T.R. Alexander, R.L. Myers, and D.P. Spangler. 1986. *The Big Cypress National Preserve.* Research Report No. 8, National Audubon Society, New York, 455 pages.

Eisenlohr, W.S., Jr. and others. 1972. *Hydrologic investigations of prairie potholes in North Dakota, 1959-68.* U.S. Geological Survey. Professional Paper 585-A. Washington D.C. 102 pages.

Fletcher, P.W. and H.W. Lull. 1963. Soil-moisture depletion by a hardwood forest during drougth years. *Soil Science Society Proceedings* 27:94-98.

Frape, S.K. and R.J. Patterson. 1981. Chemistry of interstitial water and bottom sediments as indicators of seepage patterns in Perch Lake, Chalk River, Ontario. *Limnol. Oceanogr.* 26:500-517.

Freeze, R. A. and J. A. Cherry. 1979. *Groundwater.* Prentice-Hall, Inc. Englewood Cliffs, New Jersey. 604 pages.

Freiberger, H.J. 1972. *Streamflow variation and distribution in the Big Cypress Watershed during wet and dry periods.* Bureau of Geology, Florida Department of Natural Resources. Map Series 45. Tallahassee, Florida.

Holzer, T.L. 1973. Inland wetlands and ground water in eastern Connecticut. Pages 66-83 in T. Helfgott, M.W. Lefor, and W. C. Kennard, editors. *Proceedings of First Wetlands Conference,* Institute of Water Resources Report No. 21. University of Connecticut, Storrs.

Ingram, H. A. P. 1983. Hydrology. Pages 67-158 in A. J. P. Gore, editor. *Mires: Swamp, Bog, Fen, and Moor. Ecosystems of the World 4A.* Elsevier, Amsterdam.

Junk, W. J. 1983. Ecology of swamps on the middle Amazon. Pages 269-294 in A. J. P. Gore, editor. *Mires: Swamp, Bog, Fen, and Moor. Ecosystems of the World 4B.* Elsevier, Amsterdam.

Kerfoot, O. 1968. Mist precipitation on vegetation. *Forestry Abstracts* 29:8-20.

Koryavov, P. P. 1982. Mathematical modelling of the hydrology of wetlands and shallow bodies of water. Pages 297-310 in D. O. Logofet and N. K. Luckyonov, editors. *Ecosystem Dynamics in Freshwater Wetlands and Shallow Bodies of Water, Volume 2.* Center for International Projects, Moscow, USSR.

Kulczynski, S. 1949. Peat bogs of Polesie. *Mem. Acad. Pol. Sci.* B15: 1-356.

Law, F. 1958. Measurement of rainfall, interception and evaporation losses in a plantation of Sitka spruce trees. *Proceedings of IASH General Assembly of Toronto* 2:397-411.

Leach, S.D., H. Klein, and E.R. Hampton. 1972. *Hydrologic effects of water control and management of southeastern Florida.* Florida Bureau of Geology Report Investigation 60, Tallahassee, Florida, 115 pages.

Lee, R. 1980. *Forest Hydrology.* Columbia University Press, New York, 349 pages.

Leyton, L., E.R.C. Reynolds, and F.B. Thompson. 1967. Rainfall interception in forest and moorland. Pages 163-178 in W.E. Sopper and H.W. Lull, editors. *Forest Hydrology.* Pergamon Press, Oxford.

Lichtler, W.F. and P.N. Walker. 1979. Hydrology of the Dismal Swamp, Virginia-North Carolina. Pages 140-168 in P.W. Kirk, Jr., editor. *The Great Dismal Swamp.* University Press of Virginia, Charlottesville.

Lloyd, M.G. 1961. The contribution of dew to the summer water budget of northern Idaho. *Bulletin of the American Meteorological Society* 42:672-680.

McCoy, H.J. 1974. *Summary of hydrologic conditions in Collier County, 1973.* Geological Survey, U.S. Department of the Interior. Open-File Report Fl 47030, Talahassee, Florida, 98 pages.

Millar, J.B. 1971. Shoreline-area ratio as a factor in rate of water loss from small sloughs. *J. Hydrol.* 14:259-284.

Miller, D.H. 1977. *Water at the Surface of the Earth.* Academic Press, New York, 557 pages.

Oke, T. R. 1978. *Boundary Layer Climates.* Halstead Press, New York, 372 pages.

Okruszko, H. 1967. *Hydrological factors as a basis for classification of peat bogs.* Scientific Publications Foreign Cooperation Center of Central Inst.

Scientific Technical and Economic Information, Warsaw, Poland, 13 pages.

Parker, G.G. 1974. Hydrology of the predrainage system of the Everglades in southern Florida. Pages 18-27 in P.J. Gleason, editor. *Environments of South Florida: Present and Past.* Miami Geological Society, Miami, Florida.

Pesnell, G.L. and R.T. Brown III. 1977. *The major plant communities of Lake Okeechobee, Florida, and their associated inundation characteristics as determined by gradient analysis.* South Florida Water Management District Technical Publication 77-1, West Palm Beach, Florida, 68 pages.

Romanov, V. V. 1968. *Evaporation from bogs in the European territory of the U.S.S.R.,* N. Kaner, translator. Israel Progr. Sci. Transl., Jerusalem, 183 pages.

Rutter, A.J. 1967. Evaporation in forests. *Endeavour* 26:39-43.

Spence, D. H. N. 1964. The macrophytic vegetation of freshwater locks, swamps and associated fens. Pages 306-425 in J. H. Burnett, editor. *The Vegetation of Scotland.* Oliver and Boyd, Edinburgh.

Stoeckeler, J.H. and W.R. Curtis. 1960. Soil moisture regime in southwestern Wisconsin as affected by aspect and forest type. *J. Forestry* 58:892-896.

Sturges, D.L. 1968. Evapotranspiration at a Wyoming mountain bog. *J. Soil Water Conserv.* 23:23-25.

Swank, W.T. and J.E. Douglass. 1974. Streamflow greatly reduced by converting deciduous hardwood stands to pine. *Science* 185:857-859.

Todd, D.K. 1959. *Ground Water Hydrology.* John Wiley and Sons, New York, 336 pages.

U.S. Geological Survey. 1979. *Water Summary–South Florida, March 1979.* U. S. Geological Survey, Miami, Florida, 17 pages.

Vecchioli, J., H.E. Gill, and S.M. Lang. 1962. Hydrologic role of the Great Swamp and other marshland in Upper Passiac River Basin. *J. Am. Water Works Assoc.* 54:695-701.

Ward, R.C. 1975. *Principles of Hydrology.* Second edition. McGraw-Hill Book Company (UK) Limited, Maidenhead, England, 367 pages.

Welcomme, R. L. 1979. *Fisheries Ecology of Floodplain Rivers.* Longman, London, 317 pages.

3/ A SPATIALLY DISTRIBUTED MODEL OF RAISED BOG RELIEF

G. A. Alexandrov

The cupola-like shape of a bog surface is usually explained within the framework of hydrology theory. Some propositions of this theory are used to construct a simulation model of bog growth. The model consists of two nonlinear partial differential equations. Analytical investigation has shown that there is a "smooth" convex equilibrium state. "Non-smooth" equilibrium states, which may be interpreted as the ridge-pool complex, are also available. Simulation runs are used to find initial conditions (especially bog size) leading to the ridge-pool complex.

Introduction

Ecosystem functioning in raised bogs is accompanied by intensive peat deposition. One of the consequences of this process is the leveling of the "primary" local relief together with the formation of the new specific "bog" relief (Ivanov, 1975). The central part of the bog is sometimes 10 meters higher than the margin level resulting in the cupola-like form of the bog surface (Ingram, 1982). Also characteristic are the regular structures of depressions and elevations which appear on the bog surface: the ridge-pool or the hollow-ridge complexes. Interestingly, it is peculiar to the latter to have ridges and hollows which are perpendicular to maximal surface incline—the greater the incline, the stricter the orientation. Clearly the location of the ridges and hollows follows the configuration of the contours (Romanova, 1961).

At present there are quite a number of theories to explain the origin of such a specific relief of a raised bog (Masing, 1982). However, the hydrological theory (Ivanov, 1975) seems to be most justified. Within the framework of this theory the cupola-like form of the bog derives from the higher throughflow in the marginal parts of the bog; hence their slower elevation in comparison with the central part. "When throughflow appears in the bog periphery, it retards peat accumulation here in comparison to the remaining central part. Because of this the surface of the central part of the bog gradually starts to rise above its margins, further increasing the water flow. Thus the bog relief takes a clearly pronounced cupola-like form" (Ivanov, 1975).

In general, the hydrological theory has a verbal character and its basic concepts are presented as a cause-effect scheme (causal model) in the above mentioned book. However, K. E. Ivanov introduced hydromorphological relationships—"mathematical equations linking together the water budget of bogs (determined by climatic and hydrogeological conditions in the bog site), the distribution regularities of the vegetation cover, the surface relief and the physical properties of the peat layer" (Ivanov, 1975). The bog was considered as a permanent natural formation, i.e. it was supposed that the bog relief and the physical properties of the peat layer were constant during some time interval, within which the hydromorphological relationships remained true.

In this paper, the dynamic aspect of the hydrological theory is considered, that is, the driving mechanisms of the formation process. In the next sections, the spatially-distributed model of peat accumulation is constructed, after which the steady state solution for the model equations is presented and results of simultations discussed.

Construction of the Model

According to the hydrological theory presented briefly in the introduction, the development of the bog relief is stipulated by the irregularity of peat accumulation in different parts of the bog. The local rate of this process depends mainly upon the hydrological conditions in the considered site, that is, the bog water level and throughflow. These are closely related to other factors, including the composition and structure of the vegetation cover, the temperature regime in the acrotelm and its physical properties. There is such a correlation that, through aerial photography of bog vegetation, its hydrological regime may be determined (Romanova, 1961).

Supposing that there is a complete correspondence between the distribution of vegetation and the bog water level (BWL) and throughflow, it may be thought that the rate of peat accumulation is determined only by hydrological factors. This assumption seems well grounded for natural bogs, where the variations in average long-term hydrological characteristics are a rather slow process in comparison with the induced succession of vegetation.

Throughflow is of prime importance in the verbal description for the formation mechanisms of the cupola-like bog form (Ivanov, 1975) and for the development of the ridge-pool complex (Ivanov, 1956). However, the effect of this factor seems to be less evident than that of the BWL. Therefore, it is assumed that the local intensity of peat accumulation is specified only by the local bog water level according to:

$$\partial f(x,t)/\partial t = F(f(x,t) - h(x,t)) \tag{3-1}$$

where f is the level of the bog surface and h is the level of the water table, both counted from the level of margins (see Figure 3-1).

If we think that the water motion within the peat is characterized by the hydraulic theory (Polubarinova-Kochina, 1952), then the equation for h will be

$$\partial h/\partial t = \partial/\partial x(\{\int_0^h K(f-z)dz\}\partial h/\partial x) \quad + \quad \partial/\partial y(\{\int_0^h K(f-z)dz\}\partial h/\partial y)$$

$$+ \; P \; - \; E(f\text{-}h) \tag{3-2}$$

where K is the filtration parameter, P is the precipitation, and E is the evaporation.

Equation 3-2 is the equation of fluid dynamics in the soil, which is weakly heterogeneous along the vertical and has a horizontal confining layer. Consequently, it may be used if a water confining layer runs along the line AB as shown in Figure 3-1. Since water permeability (K) of peat quickly falls with depth, the line AB may be taken as the confining layer if there is a sufficiently thick peat layer deposited above of it, i.e. if $f(t,x) \geq \mu$, where μ is such that $K(\mu) \approx 0$. Clearly the initial distribution $f(0,x)$ should also satisfy this condition. This implies that it is correct to use Equations 3-1 and 3-2 for describing the relief evolution only when the raised stage is already achieved, when a certain peat layer is deposited above the level of the margins, not only in the central part but also along the borders. However, a sufficiently thick peat layer can directly adjoin the margins only when geomorphological conditions in the site set a limit to the horizontal bog expansion. In this case, laggs occur along the junction border between the peat bed and the mineral soils (Ivanov, 1975), and the marginal slopes increase (Figure 3-1). A drainage ditch also stops the horizontal expansion of the bog, and in this sense it is an anthropogenic equivalent of the lagg. Taking into account that the lagg water level is basically determined by the watershed hydrology, it seems appropriate to specify the boundary conditions in the form:

$$h\big|_\Gamma = 0, \;\; f\big|_\Gamma = \pi \tag{3-3}$$

where, $\pi \geq \mu$, and the initial distribution set as:

$$f(0,x) \geq \mu. \tag{3-4}$$

As Equations 3-1 and 3-2 show, the hydraulic conductivity (K), total evaporation (E), and the rate of peat accumulation (F) are some functions of the bog level (BWL). The more specific form of these functions is discussed below.

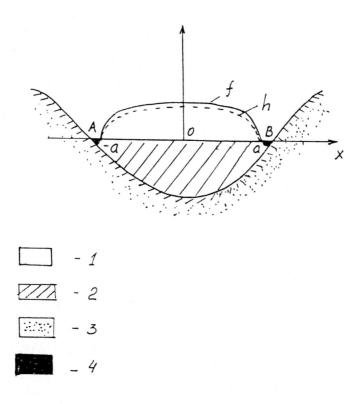

Figure 3-1. Approximate profile of a raised bog, showing (1) oligotrophic peat; (2) eutrophic peat; (3) bedrock; and (4) open water. Bog surface is **f** and water table surface is **h**.

Hydraulic conductivity

The functional relationship between the hydraulic conductivity and the BWL is well known from experiments and may be approximated either by a hyperbolic curve (Ivanov, 1957):

$$K(f-h) = B/(f-h-d)^m \qquad (3-5)$$

or by an exponent (Kozhanov, 1952):

$$K(f-h) = K_0 e^{-b(f-h)} \qquad (3-5a)$$

Evapotranspiration

The relationship between evaporation from uncovered peat and the BWL was experimentally studied by Ivitsky (1932), who suggested an exponential approximation for this relationship:

$$E(f\text{-}h) = E_0 e^{-\lambda(f\text{-}h)} \tag{3-6}$$

However, in natural conditions the decline in evapotranspiration is not that significant because of the transpiration of higher plants and the interception of rainfall by the moss cover. Therefore, as an alternative to Equation 3-6, we can consider the evapotranspiration independent of the BWL:

$$E(f\text{-}h) = E_0 = \text{constant} \tag{3-6a}$$

where E_0 is the evaporation from an open water surface.

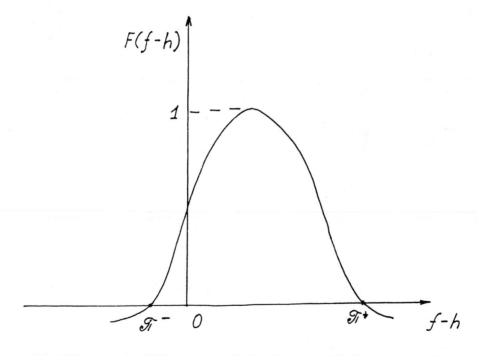

Figure 3-2. Relationship between peat accumulation rate F(f-h) and the bog water level (BWL = f-h).

Peat accumulation

Since the vertical bog growth is a rather slow process, it is hardly possible to analyze its dependence upon the bog water level (BWL) in direct experiments. At the same time it is known for certain that this process stops when the BWL is low enough (drainage) as well as in bog pools, that is, when the open water surface appears (flooding). Therefore it is natural to think that the relationship between the vertical growth rate and the BWL has the form shown in Figure 3-2 and may be approximated by the function:

$$F(f\text{-}h) = A \exp(-(f\text{-}h\text{-}\pi_0)^2 / \sigma) - \alpha \tag{3-7}$$

where $(A - \alpha)$ is the maximum rate of the vertical bog growth at "optional" BWL, which is π_0.

Form of the Bog Surface at Steady State

The concept of the steady state for a raised bog (Graulund, 1932) is discussed in detail by Frenzel (1983) and Tallis (1983). The steady state solution (f^*, h^*) of the system described in Equations 3-1, 3-2, and 3-3, which may be interpreted as the form of the bog surface and the water table at steady state, should satisfy the equations:

$$\partial/\partial x((\int_0^{h^*} K(f^*\text{-}z)dz)\partial h^*/\partial x) \; + \; \partial/\partial y((\int_0^{h^*} K(f^*\text{-}z)dz)\partial h^*/\partial y) + u^* = 0 \tag{3-8}$$

and,

$$F(f^*\text{-}h^*) = 0 \tag{3-9}$$

where $u^* = P - E(f^*\text{-}h^*)$. Function F becomes zero at the points π^+ and π^- (Figure 3-3). Consider the stationary solution, which meets the equality $f^*\text{-} h^* = \pi^+$ and $y,x \in \Omega$. Assuming that the hydraulic conductivity is given by Equation 3-5, substituting in Equation 3-8, and taking into account that $f^* = h^* + \pi^+$, we get for $m = 3$:

$$\partial/\partial x(\{ B/(2(\pi^+ + d)^2) - B/(2(h^* + \pi^+ + d)^2)\} \partial h^*/\partial x) +$$

$$\partial/\partial y(\{ B/(2(\pi^+ + d)^2) - B/(2(h^* + \pi^+ + d)^2)\} \partial h^*/\partial y) + u^* = 0 \tag{3-10}$$

Denoting

$$\varphi = h^*/(\pi^+ + d)^2 + 1/h^* + \pi^+ + d \tag{3-11}$$

it is easily seen that,

$$\partial\varphi/\partial x = \{1/((\pi^++ d)^2) \ -1/((h^* + \pi^+ + d)^2)\} \ \partial h^*/\partial x. \tag{3-12}$$

By substituting Equation 3-12 in Equation 3-10, we get

$$\partial^2\varphi/\partial x^2 + \partial^2\varphi/\partial y^2 = -2u^*/B \tag{3-13}$$

and

$$\varphi/\partial\Omega = 1/(\pi^++d). \tag{3-14}$$

Letting

$$\partial\Omega : x^2 + y^2 = a^2 \tag{3-15}$$

then after solving Equations 3-13 and 3-14 in original notations, we get

$$h^*/((\pi^++d)^2) + 1/(h^*+\pi^++d) = 2u^*/B \ (a^2-[x^2+y^2]) + 1/(\pi^++d) \tag{3-16}$$

whence,

$$h^* = (\pi^++d)^2 \ (\alpha + [\ \alpha^2 + 2 \ \alpha/(\pi^++d)]^{1/2} \ ) \tag{3-17}$$

where,

$$\alpha = u^*/B \ (a^2 - [x^2+y^2]). \tag{3-18}$$

For the hydraulic conductivity specified by Equation 3-5a, the stationary solution, which meets the equality $f^* - h^* = \pi^+$, $x,y \in \Omega$, for $\partial \Omega : x^2+y^2 = a^2$, is derived by solving the equation:

$$h^* + (1/b) \ e^{-bh^*} = (u^*/\text{æ}) \ [a^2-(x^2+y^2)] + 1/b \tag{3-19}$$

where,

$$\text{æ} = (k_0/b)e^{-b\pi^+} \tag{3-20}$$

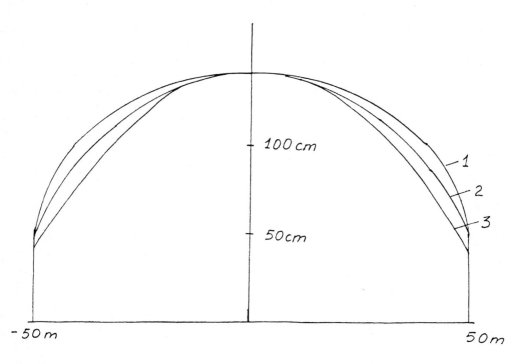

Figure 3-3. Bog profile in theory at steady state according to: 1. ellipse (Ingram, 1982); 2. Equation 3-17 (this paper); 3. parabola (Romanova, 1961).

The bog profile at steady state, as defined by the Equation 3-17, is shown in Figure 3-3. To compare, the same figure displays a parabola and an ellipse, the curves formerly suggested by Romanova (1961) and Ingram (1982) respectively for the raised bog profile.

However, the considered steady state is not unique. Since F becomes zero at two points, π^+ and π^-, there may be steady state distributions, which are combinations of parts where $f^*\text{-}h^* = \pi^+$ and $f^*\text{-}h^* = \pi^-$. Such distributions may be interpreted as hollow-ridge complexes. Simulation experiments were staged to analyze the realization conditions for steady state distributions of this kind: the initial distribution, the bog diameter, the amount of effective rainfall, and the hydraulic conductivity. Results are discussed in the next section.

Results of Simulation Experiments

According to Masing (1982), the bog relief is connected with the least ("limiting") depression diameter of the "primary" site relief. The depression borders set a limit to the bog expansion. A small limiting diameter (D) corresponds to a cupola-like raised bog with a forested center and no hollows; when D = 0.6 km, a raised bog is usually formed with forested margins and small hollows in the deforested center, and so on (Masing, 1982).

We tried to reproduce this phenomenon within the framework of simulation experiments. The hydraulic conductivity was given by the Equation 3-5a and the evaporation by Equation 3-6a. Figure 3-4 presents the form of the initial distribution. Simulation experiments were staged for D = 100 m and D=500 m. Their aim was to find the value of u^* which would provide a smoothing in the first case (D=100m) and would further the differentiation of initial relief heterogeneities in the second case (for the same value u^*). Figures 3-5 and 3-6 show that such a value exists and equals 0.5 cm/year (K = 9.83 cm/sec and b = 0.25 cm^{-1}). It could be perhaps higher for another choice of K and b, since, in Equation 3-17, the height of the water table is related to the value u^*b/k_0. It should be noted that for the differentiation of the relief in the case D = 100 m, we need an essentially higher amount of effective rainfall, $u^* = 30$ cm/year, while in the case D = 500 m already for $u^* = 1.5$ cm/year, starting from some time, the level of the water table, h, exceeds the surface level, f, in the central part.

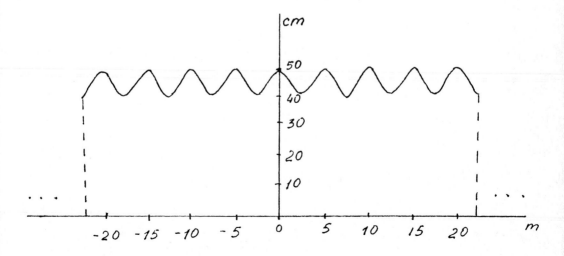

Figure 3-4. The initial bog profile used in simulation experiments.

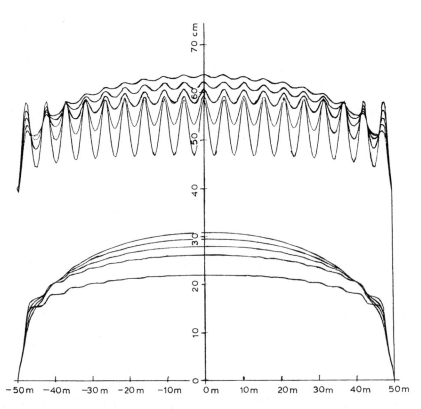

Figure 3-5. Variation in the bog profile in computer runs with diameter (D) equal to 100 meters.

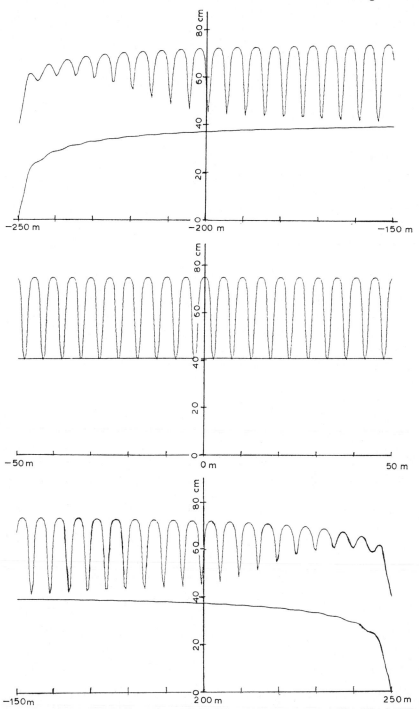

Figure 3-6. The final bog profile in computer runs with diameter (D) equal to 500 meters. Only parts of the profile are shown.

Conclusion and Discussion

In interpreting the results of simulation experiments, one may conclude the following:

1. The relationship between the bog relief and its diameter may be explained in terms of hydrology.

2. There is no need to take into account the effect of throughflow upon the local rate of peat accumulation; it is enough to take account only of the bog water level.

3) The analysis of the relationship between the bog height and its diameter, based on rich empirical material (300 raised bogs), showed Granlund (1932) that there exists a limiting height of the bog dome, which is never exceeded even in the largest bogs The existence of this limiting height of the bog dome may be stipulated by the fact that as the diameter grows, steady state, such as in Equation 3-17, is attained at lower values of u^*. On the other hand, the steady state of the ridge-hollow-complex type needs a smaller height of the bog dome for the outflow of the same amount of effective rainfall, due to a higher water flow in the hollows. However, within the framework of the simulation experiments, it was assumed that the site geomorphological conditions set a limit to the horizontal expansion of the bog, i.e. we considered a rather special case, since the vertical bog growth, especially at the initial stages, is usually accompanied by horizontal expansion. Perhaps by taking into account this process we could reach more meaningful conclusions about the dynamics of relief formation, especially for D = 500 m. Nevertheless, the formal description of the process of horizontal expansion promises to be more complicated than the above suggested model due to the inconsistency of the hypothesis that the water confining level is horizontal. It should be noted that we need experiments for the two-dimensional case, when applying the model to a concrete object, for instance, to interpret the stratigraphy of the peat deposit for a more adequate climate reconstruction. This case needs no model modifications and the only limit, after the appropriate computation scheme is chosen, is set by the available CPU time and on-line storage.

References

Frenzel, B. 1983. Mire—repositories of climatic information or self perpetuating ecosystems? Pages 35-59 in A.J.P. Gore, editor. *Mires: Swamp, Bog, Fen and Moor. Ecosystems of the World, Vol 4A.* Elsevier, Amsterdam.

Granlund, E. 1932. De svenska hogmossarnas geologi. *Sver. Geol. Unders. Ser. C* 373:1-993.

Ingram, H.A.P. 1982. Size and shape in raised mire ecosystems: a geophysical model. *Nature* 297:300-303.

Ivanov, K.E. 1956. Formation of ridge-hollow relief as a result of moisture outflow from bogs. *Bull. of Leningrad Univ., Ser. Geological and Geographical* 12:58-72 (In Russian).

Ivanov, K.E. 1957. *Foundations of Hydrology of the Forest Zone Bogs.* Hydrometeoizdat, Leningrad, 500 pages (In Russian).

Ivanov, K.E. 1975. *Water Exchange in Boggy Landscapes.* Hydrometeoizdat, Leningrad, 280 pages (In Russian).

Ivitsky, A.I. 1932. Evaporation from peat soil as function of the climatic factors and bog water level. *Pochvovedenie* 1932(2):267-283 (In Russian).

Kozhanov, K. Ya. 1952. *Investigation of Distance Between Drains in Agricultural Utilization of Bogs in BelSSR.* BelSSR Academy of Sciences Publishing House, Minsk, 102 pages (In Russian).

Masing, V.V. 1982. The dynamics of bog ecosystems and water bodies. Pages 22-36 in D.O. Logofet and N.K. Luckyanov, editors. *Ecosystem Dynamics in Freshwater Wetlands and Shallow Water Bodies, Vol 1.* Center for International Projects, Moscow, USSR.

Polubarinova-Kochina, P. Ya. 1952. *Theory of Groundwater Motion.* Gostechizdat, Moscow, 643 pages (In Russian).

Romanova, E.A. 1961. *Geobotanical Foundations of Hydrological Studies in Raised Bogs.* Hydrometeoizdat, Leningrad, 244 pages (In Russian).

Tallis, J.H. 1983. Changes in wetland communities. Pages 311-344 in A.J.P. Gore, editor. *Mires: Swamp, Bog, Fen and Moor. Ecosystems of the world, Vol. 4A.* Elsevier, Amsterdam.

4/ INTERFERENCE BETWEEN MOSSES AND TREES IN THE FRAMEWORK OF A DYNAMIC MODEL OF CARBON AND NITROGEN CYCLING IN A MESOTROPHIC BOG ECOSYSTEM

D.O. Logofet
G.A. Alexandrov

Based upon a complete balance scheme for the organic matter and nitrogen cycles through the ecosystem of a meso-trophic (transitional) bog, a series of mathematical models is developed to see those trends in the evolution of the ecosystem which are caused by quantitative regularities of cycling. The most advanced, dynamic model of that séries describes cycling through five aggregated compartments (trees, dwarf shrubs, grasses, mosses, and litter) and takes into account, along with competition for mineral nutrients, these two phenomena: (1) a weakness of plants under nitrogen starvation, and (2) an increase in litter decomposition in response to an increase in the nitrogen content of dead organic matter. Inclusion of the interference effect between trees and mosses in the model has not changed the principal conclusion that the tendency in evolu- tion of the given transitional bog is towards a forest phase.

Introduction

One can distinguish at least three types of models for matter cycling through an ecosystem: conceptual, balance, and simulation. The conceptual models of matter cycling are usually represented in the form of flow diagrams and frequently called compartmental models, the compartments standing for different components of an ecosystem. In the case where numerical values of storages in the compartments and flows among them are known, the models are referred to as balance ones. Elaboration of a simulation model, as a rule, relies upon the conceptual and balance ones, some speculations or hypotheses being used about the factors which regulate intensity of the flows. Such a combination of a conceptual model with the corresponding factors is sometimes called a causal model.

To adjust a simulation model for a particular ecosystem it is necessary to have a balance model of that ecosystem. In other words, investigation of an ecosystem by means of simulation modelling should usually begin with elaboration of a conceptual model, then a balance model, and then a causal one. This sometimes requires a fairly long period of field observations and laboratory measurements.

In terrestrial ecosystems, such a problem has been mostly solved for a few examples only—for the organic matter turnover (Dauffenbach et al., 1981; Bunnell and Scoular, 1975), nitrogen turnover (Rosswall et al., 1975), phosphorus turnover (Bunnell, 1971), and the water cycle for tundra ecosystems, in the framework of IBP (Miller et al., 1975). The same substance turnovers have been analyzed by means of simulation modelling for cypress wetland ecosystems (see Hall and Day, 1977, and references therein) and for the North American marshes (Bayley and Odum, 1976; Mitsch et al., 1982). The research efforts in the field of nitrogen and phosphorus turnover are explainable from the recognition of the importance of these elements in the functioning of plant communities.

At the same time, application of potassium fertilizers in a raised mire, when used for forest-growing, has revealed an important role of potassium in the functioning of the mire ecosystem. Since the turnovers of all these elements in a biogeocoenosis are related in some sense, via stoichiometric ratios for living matter of plants, it is worth considering a combined turnover of the most important elements. Such a consideration, however, has been constrained by the lack of balance models for a number of elements. Therefore, in the model discussed below, a combined turnover of only two elements, carbon and nitrogen, is considered, though the methodology used can be extended for the case of an arbitrary number of elements.

The Dynamic Model

The choice of variables for a model is almost universally recognized as one of the most nonformal phases in the course of constructing the model. Nevertheless, such a formal procedure as the environ analysis (Patten, 1982; see also Chapter 11) can give some quantitative characteristics for the role of an ecosystem component in matter cycling through the ecosystem, thus facilitating a choice of variables. Though animals (phytophages and saprophages) and soil microorganisms are among the main components of any ecosystem, separating them into individual state variables requires introduction of both additional hypotheses and the model parameters which can hardly be estimated. When modelling tundra ecosystems, for example, the animals and microorganisms are usually considered as state variables, whereas this has not been the case in ecosystem modelling of cypress wetlands and marshes. As we have seen from the environ analysis of the organic matter turnover through a mesotrophic bog ecosystem (Logofet and Alexandrov, 1984a), the role of animals is relatively low in that ecosystem, while the soil microorganisms (fungi and bacteria) are related to the litter so closely that it seems unreasonable to consider them as separate state variables. Therefore, the "small turnover" of organic matter (Walter, 1979) was investigated first in the case study of the bog ecosystem.

The mean-annual hydrological conditions (i.e. the groundwater level and the flowing capacity) were supposed invariable in the sense that they had neither significant variations from year to year nor significant trends in the long term (the hypothesis of invariance in hydrological regime). Any alterations in organic matter cycling were thought to be induced by some alterations in the nitrogen cycle. Their close dynamic interrelation is provided by two simple mechanisms: (1) decreasing the N/C (nitrogen/carbon) ratio in LOM (Living Organic Matter) results in an increase in total litterfall (weakness of plants under a nitrogen starvation; see Larcher, 1975); (2) increasing the N/C ratio in DOM (Dead Organic Matter) results in an increase in a DOM decomposition rate (Walter, 1979). The dynamic interactions between the simplest cycles of organic matter and nitrogen are shown in Figure 4-1 in the form of a systems diagram.

The simplest way to formalize the above mechanisms (by means of linear dependences on carbon/nitrogen (C/N) or nitrogen/carbon (N/C) ratios results in a system of nonlinear differential equations (Logofet and Alexandrov, 1984b):

$$dC_1/dt = -O_C C_1{}^2/N_1 + A_C C_1 \tag{4-1}$$

$$dN_1/dt = -O_N C_1 + (1-c)D_N N_2{}^2/C_2 + A_N \tag{4-2}$$

$$dC_2/dt = O_C C_1{}^2/N_1 - D_C N_2 - f_C C_2 - T_C C_2 + V_C \tag{4-3}$$

$$dN_2/dt = O_N C_1 - D_N N_2{}^2/C_2 - T_N N_2 + V_N \tag{4-4}$$

Mechanisms (1) and (2) are reflected in the dynamic model by the terms $O_C C_1{}^2/N_1$ and $D_N N_2{}^2/C_2$, while the rest of the terms in the equation have the same meaning as in Figure 4-1. The coefficients of the equations and the initial values of their variables were estimated according to the conceptual-balance model of chemical element cycling through a mesotrophic bog ecosystem (the transitional rim of a raised bog) within Tayozhny Log watershed, located in the southern taiga subzone, Valdaj District, Novgorod Region, USSR (Bazilevich and Tishkov, 1982).

The simulation runs have shown that the system comes to an equilibrium state possessing approximately the same storages in LOM and DOM and the

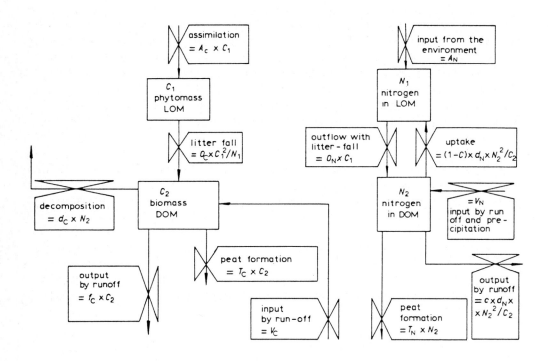

Figure 4-1. Dynamic model for the organic matter turnover depending on the nitrogen cycle in a mesotrophic bog (from Logofet and Alexandrov, 1984a). In the present paper, somewhat different notation is used (e.g. O_C in this figure is O_iC in the text).

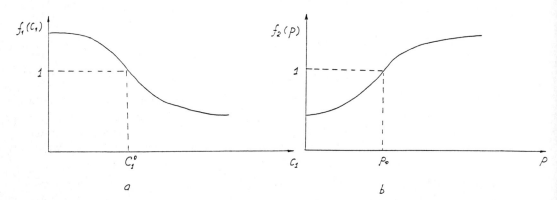

Figure 4-2. Qualitative form of interferences between mosses and trees, showing a) dependence of moss biomass growth on shading, and b) dependence of plant die-off on the rate of *Sphagnum* moss growth.

same flows among them as those in the initial state. The presence of such an equilibrium state seems to be nothing more than a reflection of the balanced nature of the matter cycling scheme and a consequence of the lack of any long-term trends in the environmental conditions.

However, the inclination to describe ecosystem dynamics in some more meaningful terms has led to division of the LOM component into several smaller subcomponents, namely trees, dwarf-shrubs, grasses, and mosses. Moreover, as the main phytocenosis subcomponents, they differ in both their N/C values and the mechanisms for sustaining those values at a level that is optimal for a given vegetation type. Therefore, not only is the subdivision of LOM necessary, but it is also necessary to provide different formalizations of the sustaining mechanisms for specific subcomponents as well (Logofet and Alexandrov, 1984b).

Combined with some stability considerations, those different formalizations yielded the equations for the dynamic model, accounting also for competition among the components for mineral nutrients (nitrogen) (Logofet and Alexandrov, 1984b):

$$dC_1/dt = -O_1{}^C C_1{}^2/N_1 + A_1{}^C C_1 \qquad (4\text{-}5)$$

$$dN_1/dt = -O_1{}^N N_1 + r_1 C_1 s/(\ r_1 C_1 + r_2 C_2 + r_3 C_3 + r_4 C_4) + V_1{}^N \qquad (4\text{-}6)$$

$$dC_2/dt = -O_2{}^C C_2{}^2/N_2 + A_2{}^C N_2 \qquad (4\text{-}7)$$

$$dN_2/dt = -O_2{}^N N_2 + r_2 C_2 s/(\ r_1 C_1 + r_2 C_2 + r_3 C_3 + r_4 C_4) + V_2{}^N \qquad (4\text{-}8)$$

$$dC_3/dt = -O_3{}^C C_3{}^2/N_3 + A_3{}^C N_3 \qquad (4\text{-}9)$$

$$dN_3/dt = -O_3{}^N C_3 + r_3 C_3 s/(\ r_1 C_1 + r_2 C_2 + r_3 C_3 + r_4 C_4) \qquad (4\text{-}10)$$

$$dC_4/dt = -O_4{}^C C_4{}^2/N_4 + A_4{}^C C_4 \qquad (4\text{-}11)$$

$$dN_4/dt = -O_4{}^N C_4 + r_4 C_4 s/(\ r_1 C_1 + r_2 C_2 + r_3 C_3 + r_4 C_4) + V_4{}^N \qquad (4\text{-}12)$$

$$dC_5/dt = \sum_{i=1}^{4} (O_i{}^C C_i{}^2/N_i) - D_C N_5 - T_C C_5 + V_5{}^C \qquad (4\text{-}13)$$

$$dN_5/dt = (O_1{}^N N_1 + O_2{}^N N_2 + O_3{}^N N_3 + O_4{}^N N_4) - D_N N_5{}^2/C_5$$
$$- T_N N_5 + V_5{}^N \tag{4-14}$$

$$s(t) = (1-c)D_N N_5{}^2(t)/C_5(t) \tag{4-15}$$

Here C_i and $N_i (i = 1,...,5)$ designate, respectively, the organic matter and nitrogen stocks in trees, dwarf shrubs, grasses, mosses, and litter including the 15-cm top surface layer of peat. $V_i{}^N$ are nitrogen inflows from the outside, into trees and dwarf shrubs from peat, into mosses from precipitation, into the litter from precipitation, and from adjacent ecosystems from runoff; s is the total nitrogen uptake by phytocenosis; c is a runoff coefficient (the proportion of nitrogen carried out of the litter by runoff); r_i are competition coefficients; $O_i{}^C$, $O_i{}^N$, $A_i{}^C$, D_C, D_N, T_C, and T_N are constant quantities having equivalent meaning as in Figure 4-1, except for the index i.

Table 4-1. Production and biomass of the plant components of a mesotrophic bog.

Components	Simulation experiment[a]			
	1	2	3	4
Trees, C_1				
Production, g/m²·yr	358.0	71.6	89.3	17.5
Biomass, g/m²	18,000	3,600	4,488	879
Dwarf shrubs, C_2				
Production, g/m²·yr	110.1	31.3	113.2	113.2
Biomass, g/m²	704	200	724	724
Mosses, C_4				
Production, g/m²·yr	33.0	19.1	33.2	33.2
Biomass, g/m²	215	125	216.8	216.8

[a]Simulations: 1–equilibrium state in the non-modified model; 2–"quasi-stationary" state in the non-modified model; 3–unstable equilibrium in the modified model; 4–stable equilibrium in the modified model.

It was shown that the model has a stationary state which is "feasible" under a sufficient level of mineral nutrition. For the particular ecosystem modelled, this stationary state was interpreted (according to biomasses of the plant components and to values of their production flows) as a forest phytocenosis. Since the area under study represents an ecotone zone separating a forest from a raised bog, one can conclude that the present hydrological conditions promote the "forest attack" to the bog. The reverse process, namely, the "bog attack" to the forest, can proceed in nature as well, without any change in the hydrological regime but with an appropriate change in the mineral nutrition regime (oligotrophication). We cannot say, however, that the model, in its original form, is capable of reproducing this phenomenon: the decrease in the nitrogen inflow leads to infeasibility of the stationary state. The corresponding simulation experiments show that the system transits into some "quasi-stationary" state which, according to its biomass stock in trees, can be identified with a raised bog. The relation between the trees' and mosses' biomasses are similar to that of the transitional bog (see Table 4-1).

In order to have a more adequate description for the course of the "bog attack" to the forest, we attempt, in the present paper, to include into the model an effect of a direct interference between mosses and trees along with competition for nitrogen (among all the components) and with the stoichiometry-sustaining mechanisms.

Interference Between Mosses and Trees

Interference, as an ecological interaction, means a mutual oppression distinct from competition for common resources. The interactions of such a kind are often used to explain succession processes. For example, *Sphagnum* mosses, as light-requiring plants, are very sensitive in their rates of growth to shade or light. Hence the term $A_4^C C_4$, corresponding to growth of mosses' biomass in Equation 4-11, should be modified by a function, $f_1(C_1)$, reflecting a dependence of biomass growth upon a level of shading, which, in turn is determined in the model by tree biomass. From general reasons the function must have the form shown in Figure 4-2. (When there are practically no trees, the mosses' growth attains some maximum level from which it declines steadily in response to increase in shading). In order to promote analytical investigations, the function $f_1(C_1)$ is supposed further to admit an approximation by $(C_1^0/C_1)^\alpha$ with some parameter $\alpha > 0$ in the vicinity of the initial value C_1^0 of tree biomass estimated in accordance with the balance scheme by Bazilevich and Tishkov (1982).

There is another substantial mechanism in the succession of a bog ecosystem, caused by the property of *Sphagnum* to grow up permanently from year to year as its lower parts die off and turn into peat. "It is clear that only those plants can successfully develop on a *Sphagnum* cover which possess an ability to follow the *Sphagnum* up, not lagging behind, and thus avoiding the risk of being buried. On the contrary, those plants which have no such ability are doomed to poor vegetation in a mire" (Sukhachev, 1973). So, to include this phenomenon in the model we should modify the terms of Equations 4-5 and 4-6 that represent the litterfall ($O_1{}^{CC}C_1{}^2/N_1$) and the nitrogen outflow with litterfall ($O_1{}^{NN_1}$), by a function $f_2(p)$ where p is the current rate of *Sphagnum* growth, $p = A_4{}^{CC}C_4f_1(C_1)$. The function $f_2(p)$ must have the form shown in Figure 4-2b (starting from a non-zero point since the litter-fall is not zero even if the mosses are completely absent). In the vicinity of the point p0 representing the initial estimated value of p we approximate $f_2(p)$ by a function:

$$(p/p_0) = (C_4/C_4{}^0)\,(C_1{}^0/C_1)^\alpha. \tag{4-16}$$

The equilibrium values $C_2{}^*$, $C_3{}^*$, $C_4{}^*$ of the modified model are determined by the same expressions as before (Logofet and Alexandrov, 1984b), while $C_1{}^*$ is to be found from the equation:

$$-(O_1{}^N\,O_1{}^C/A_1{}^C)\,C_1{}^*\,(C_4{}^*/C_4{}^0)(C_1{}^0/C_1{}^*)^{2\alpha} + r_1 C_1/\beta_3 + V_1{}^N = 0 \tag{4-17}$$

whose general form is the following:

$$-ay^{1-2\alpha} + by + c = 0 \tag{4-18}$$

The numerical analysis of stability by the Monte-Carlo technique has shown that the equilibrium state is stable in Liapunov sense when $\alpha < 1/2$. In what follows we consider $\alpha = 1/4$ as a typical value of $\alpha < 1/2$. Then, two equilibrium values of C_1 are possible, namely:

$$(C_1{}^*)_1 = 4{,}488 \ g/m^2 \tag{4-19}$$

and,

$$(C_1{}^*)_2 = 879 \ g/m^2 \tag{4-20}$$

These values can be interpreted as a forested bog and an open bog respectively. As shown by simulation, the former is apparently unstable (the trajectories go either toward the second equilibrium, Figure 4-3, or toward infinity, Figure 4-4, where the approximation is no longer valid). As can be seen from silulation runs, the second equilibrium state is stable under a sufficient level of mineral nutrition (s > 1,365 g/m^2.yr) and the trajectories converge if initiated from $C_1(0)$ < 4,488 g/m^2 (Figure 4-3). Note that the estimated initial state does not belong to the attraction domain of the second equilibrium state.

Discussion and Conclusions

The following implications can be drawn from the above results:

1. Accounting for interference between mosses and trees, the previous conclusions about the tendency in evolution of the modelled bog to a transition into a forest phase (Logofet and Alexandrov, 1984 b) are preserved.

2. A transition into an open raised bog phase is possible only after a catastrophic reduction of stand biomass almost two times below its present level, which could happen, for example, by fire or flooding.

It should be noted, however, that we are far from accounting for all possible mechanisms of interference among plants. In a similar way one could take into account the interference between shrubs and mosses, or between grasses and mosses. As a result, the number of equilibrium states should increase. For instance, two equilibrium states may turn out possible which could be interpreted as open raised bogs having the same biomass stocks of trees, but different ones of shrubs. In view of the great diversity of mire phytocenoses in nature, this multi-equilibrium character of the model would allow the interpretation of its results to be more meaningful in the botanical sense. The present conclusions, however, about the tendency in evolution of the given bog seem to remain valid.

The more complicated problem is that of invariance of the hydrological regime. The invariance hypothesis strongly restricts the range of practical applications of the present model, which at present can be referred to as hypothetical and theoretical, rather than management-oriented. Discounting this hypothesis can greatly affect the results of the model research, as shown by Logofet and Alexandrov (1984b). Taking the hydrological factors into consideration is possible under conjunction with an appropriate hydrological model (e.g. Alexandrov et al., 1983), although it requires a considerable amount of data on hydrology for the area under study.

In some cases, however, the hypothesis of water regime invariance can be adopted with a sufficient level of confidence. A project of growing a forest in a raised mire with controlled drainage by application of mineral fertilizers can be referred in such a case.

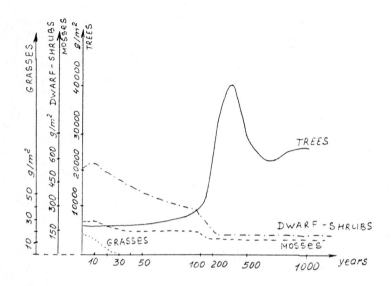

Figure 4-3. Simulation run for mesotrophic bog model with the initial state and flows corresponding to the balance model of Bazilevich and Tishkov (1982).

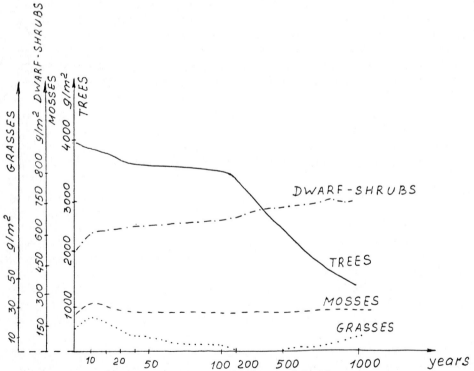

Figure 4-4. Simulation run for mesotrophic bog model with the initial state of tree biomass less than 4,488 g/m^2.

References

Alexandrov, G.A., L.L. Golubyatnikov, and T.E. Shytikova. 1983. Model of a water regime in a raised bog. Pages 12-13 in G.V. Vorapaev et al., editors. Proceedings All-Union Conference "Problemy okhrany, izchenia i ratsionalnogo ispolzovaniya vodnykh resursov" (*Issues of Conservation, Research and Rational Use of Water Resources*). Inst. for Water Research, USSR Academy of Science, Moscow. (in Russian)

Bayley, S., and H.T. Odum. 1976. Simulation of interrelations of the Everglades marsh, peat, fire, water, and phosphorus. *Ecol. Modelling* 2:169-188.

Bazilevich, N.T., and A.A. Tishkov. 1982. Conceptual balance model of chemical cycles in a mesotrophic bog ecosystem. Pages 236-272 in D.O. Logofet and N.K. Luckyanov, editors. *Ecosystem Dynamics in Freshwater Wetlands and Shallow Water Bodies, Vol 2*. Centre of International Projects, Moscow, USSR.

Bunnell, F.L:, 1971. TCOMP—an interactive compartment model of nutrient flux and decomposition. Pages 17-29 in J. Brown, coordinator. *The Structure and Function of the Tundra Ecosystem,* U. S. Tundra Biome 1971 Progress Report, Vol. 1, pages 17-29.

Bunnell, F.L., and Scoular K.A., 1975. ABISCO II: a computer simulation model of carbon flux in tundra ecosystems. Pages 425-448 in T. Rosswall and O.W. Heal, editors. *Structure and Function of Tundra Ecosystems*. Ecol. Bull. (Stockholm, Sweden), Vol. 20.

Dauffenbach, L., W.J. Mitsch, and R.M. Atlas. 1981. Modelling the fate of crude petroleum spills in Arctic tundra ecosystems. Pages 893-916 in D.M. Dubois, editor. *Progress in Ecological Engineering and Management by Mathematical Modelling*. Proceedings of the Conference. Centre Belge d'Etudes et de Documentation, Liège, Belgium.

Hall, C.A.S. and J.W. Day, Jr. 1977. *Ecosystem Modelling in Theory and Practice*. Wiley Interscience, New York, 689 pages.

Larcher, V. 1975. *Physiological Plant Ecology*. Springer-Verlag, New York, 252 pages.

Logofet, D.O. and G.A. Alexandrov. 1984a. Modelling of matter cycle in a mesotrophic bog ecosystem. I. Linear analysis of carbon environs. *Ecol. Modelling* 21:247-258.

Logofet, D.O. and G.A. Alexandrov. 1984b. Modelling of matter cycle in a mesotrophic bog ecosystem. II. Dynamic model and ecological succession. *Ecol. Modelling* 21:259-276.

Miller, R.C., B.D. Collier, and F. Bunnell. 1975. Development of ecosystem modelling in the Tundra Biome. Pages 95-115 in B.C. Patten, editor. *Systems Analysis and Simulation in Ecology, Vol. III*. Academic Press, New York.

Mitsch, W.J., J.W. Day, Jr., J.R. Taylor and C. Madden. 1982. Models of North American freshwater wetlands. *Int. J. Ecol. Environ. Sci.* 8:109-140.

Patten, B.C. 1982. Environs: relativistic elementary particles for ecology. *Amer. Nat.* 119:179-219.

Rosswall, T., J.G.K. Hower-Ellis, L.G. Johansson, S. Jonsson, B.E. Ryden, and M. Sonesson. 1975. Stordalen (Abisko), Sweden. Pages 264-294 in T. Rosswall, and O.W. Heal, editors. *Structure and Function of Tundra Ecosystems.* Ecol. Bull. (Stockholm, Sweden), Vol 20.

Sukhachev, V.N. 1973. *Izbrannye Trudy* (Selected Works), Vol. 2., Nauka, Leningrad, pages 88-89. (In Russian)

Walter, H. 1979. *Allgemeine Geobotanik.* Eugen Ulmer, Stuttgart, 260 pages.

5/ SIMULATION MODELS OF COASTAL WETLAND AND ESTUARINE SYSTEMS: REALIZATION OF GOALS

Charles S. Hopkinson
Richard L. Wetzel
John W. Day, Jr.

The most successful models of marsh/estuarine systems are more than an academic exercise in summarizing large data sets. They are also useful tools for formulating new testable hypotheses, for guiding large ecosystem-level research programs and for guiding the management of coastal habitats. Reviewed here are six simulation models of coastal marsh/estuarine systems that are representative of both management- and research-directed efforts. The authors exemplify the diversity of objectives, approaches and usefulness of these models. The utility of sensitivity analysis as a means to reveal factors controlling certain ecosystem behaviors is analyzed and shown to be strongly limited by the initial abstraction or conceptualization of the ecosystem structure.

Introduction

Simulation modelling of ecosystems has progressed substantially over the last 15 years; we now have a clearer understanding of our models and are better able to apply models to the specific questions. Advances in ecological modelling are in part due to improvements in computer technology and to the demand for integrative syntheses of ecosystem-level studies that involve multidisciplinary teams of scientists. Simulation models are far more than tools for synthesis and integration of hierarchical level data, however. Models are now used to plan and guide research programs from the outset—to guide research programs, to identify data weaknesses and gaps, to evaluate management-oriented alternatives, and, perhaps most importantly, to formulate testable hypotheses. As ecological modelling matured and became a valuable tool for ecological research, there was less need for modellers to "oversell" or promise too much in the way of predictive output to their colleagues and funding agencies. Predictions tend to be much more conservative today.

Although there are literally hundreds of published mathematical models in ecology, relatively few are of coastal marsh and estuarine ecosystems. This is in spite of the fact that estuarine salt marshes are one of the most frequently studied of all ecological systems (Summers et al., 1980). Models of coastal systems tend to be less complex relative to their terrestrial system counterparts. They are usually characterized by having few and highly aggregated state variables. Most are trophically designed and use energy or carbon as the conserved unit of exchange. Control functions are usually rather rudimentary in nature due primarily to a lack of available information describing the exact nature of control.

This paper attempts to exemplify the diversity of model objectives, approaches, and product utility by comparing and contrasting six published simulation models of the coastal marsh/estuarine ecosystem. We present our ideas concerning where ecosystem modelling is progressing and some experimental model manipulations that could promote understanding of models and make them more useful in guiding research and management efforts in coastal systems.

Choice of Models

Six simulation models of coastal marsh/estuarine systems were chosen to be representative of both management and research-directed efforts. All had the objective of being more than an academic exercise of summarizing large and complex data sets. These models illustrate the importance of modelling as a tool for understanding broader objectives such as guiding research through formulation of testable hypotheses and in guiding the management of coastal habitats. Two modelling programs described are clearly management-oriented and four are concerned primarily with basic research. The basic research-directed models portray the strong and intimate role that modelling has assumed in large ecosystem-level research programs. These are models that were initialized early in research program design and to a large extent have evolved and developed in the course of research program building.

The first of the six models exemplified actually consists of a group of both conceptual and simulated models (Costanza et al., 1983; Leibowitz and Costanza, 1983; Fruci et al., 1983) of the Mississippi River Deltaic Plain region in Louisiana, USA. The series is strongly management-oriented and attempts to pull together information from several hierarchical levels ranging from the salt marsh habitat to the entire southern Mississippi River drainage basin, including urban centers. Hopkinson and Day (1980a,b) used a series of models to evaluate effects of urbanization on nutrient runoff to a coastal swamp forest and to evaluate various hydraulic options for promoting water runoff and decreasing lake eutrophication. Two long-term and evolving modelling programs (Wiegert et al., 1975; Wiegert and Wetzel, 1979; and Summers et al., 1980; Summers and McKellar, 1981) deal with the salt marsh ecosystem. These two programs have many similarities but at the same time strongly contrast each other in their philosophy of model control. In one, a great deal of attention is directed at "biologically" realistic controls at the population level while the other accentuates control by the major physical driving forces that seem to be basic to the functioning of the salt marsh ecosystem. The role of microbes and the importance of detritus in energy flow of the aquatic portion of a salt marsh system is addressed in the series of models by Christian and Wetzel (1978) and Wetzel and Christian (1984). Vorosmarty et al. (1983) described a model similar to that used by Hopkinson and Day (1980b) but with a totally different set of objectives. Their interest was in utilization of the model not only in planning and guiding research but also in developing research hypotheses.

Conceptual Models of the Mississippi River Deltaic Plain Region

In response to a request from the U.S. Fish and Wildlife Service, a series of hierarchical, conceptual models were developed and quantified for the Mississippi River Deltaic Plain Region (MDPR-Costanza et al., 1983; Bahr et al., 1983). The purpose of these models was to summarize existing data for the MDPR in a form useful to scientists and coastal managers. The models integrate information on ecology, hydrology, climatology, and socioeconomics of 20 ecological and economic habitats and seven hydrologic units into which the MDPR has been divided. These models are a continuation and product of several decades of research by the Center for Wetland Resources at Louisiana State University and were designed for understanding and management of this region. The MDPR of the Louisiana and Mississippi coastal zones consists of the broad, topographically flat area that includes the largest active delta system in North America. The region's dynamic nature, high biological productivity, and intense level of economic activity have combined to create resource management problems of enormous magnitude.

Conceptual models were constructed in a hierarchical manner for several purposes: 1) to organize existing information on relevant temporal and spatial scales; 2) to organize environmental and management problems within this framework; and 3) to target areas in need of additional research.

Hierarchically, the region was abstracted into three spatial scales. Considered in order of increasing size were habitats, hydrologic units, and the entire region (Figure. 5-1). The conceptual models describe the overall compartmental and flow structure and interactions and forcing functions considered most important. To facilitate information retrieval by management agencies and prepare for possible future simulation studies, data were further organized and tabulated in input-output format. The forcing functions included at the various levels of model conceptualization represent hypothetical statements of the

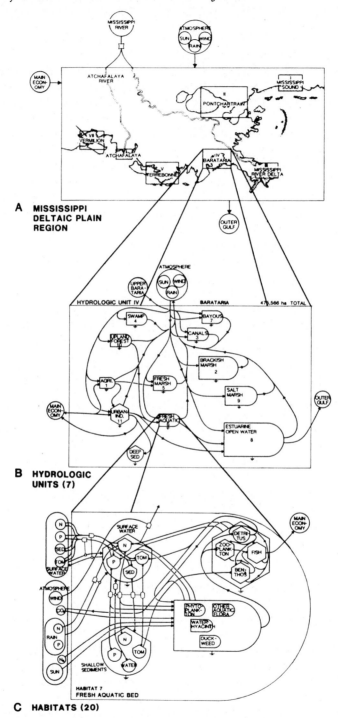

Figure 5-1. Conceptual models of hierarchical structure of the Mississippi River Deltaic Plain Region (from Costanza et al., 1983).

controlling role of external physical factors.

At the highest hierarchical level—the region—the most important forcing functions are riverine input (especially the Mississippi River), the Gulf of Mexico, and the atmosphere (Figure 5-2). It is the interaction of sediment deposition by the river and erosional processes of the Gulf of Mexico which have largely shaped the overall region, both economically and geologically, over the past several thousands years.

At the level of the hydrologic unit, water flow and interaction between different habitats are the most obvious elements of compartmental function and structure (Figure 5-3). At this level of organization, data began to be added to the models. A simulation model was developed to estimate the flows of water, nutrients, and organic matter within the hydrologic units.

Finally, ecological structure and function became most apparent at the level of the habitat. An example of this is the conceptual diagram for the cypress-tupelo swamp forest habitat (Fig. 5-4). The most important forcing functions controlling internal carbon/energy flows were surface water input and atmospheric parameters (e.g., sunlight and precipitation). Spatial heterogeneity within the swamp system was accomodated in the model structure by dividing the region into four subcompartments: 1) sediments and associated water and nutrients; 2) surface water and associated nutrients; 3) vegetation and litter; and 4) heterotrophs. As conceptualized in this basic format, energy flows are only donor-controlled with the exception of primary productivity which is controlled by a combination of water and nutrient availability. Standing stock and flow data for selected habitats were systematically organized in input-output tables. The combination of the flow diagram and the input-output table allows efficient transfer of information regarding the structure and interactions of the habitat as well as a coherent and organized fashion presentation.

One of the main purposes of the modelling effort was to provide information about impact and management at a variety of spatial levels. Listed below are some representative issues spatially grouped.

Regional Level

1. *Wetland Loss*: The conversion of wetland habitats to open water is accelerating in much of the MDPR. The present rate is estimated at 100 km²/yr (Gagliano et al., 1981). This loss is the result of interrelated natural and cultural processes, including sea level rise, erosion from dredging operations, Mississippi River diversion, subsidence, saltwater intrusion, and the sediment starvation of marshlands.

2. *River Switching*: The Atchafalaya River is currently poised to divert much of the flow of the Mississippi River, with potentially major consequences for the economic structure of the region.

3. *Industrial Pollution*: Water and air quality and chemical waste disposal are major issues in the MDPR, which includes petrochemical and port facilities that are among the most active in the world.

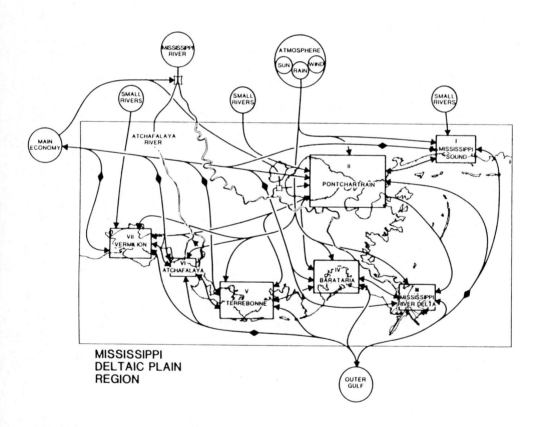

Figure 5-2. The regional perspective of the Mississippi River Deltaic Plain Region (from Costanza et al., 1983).

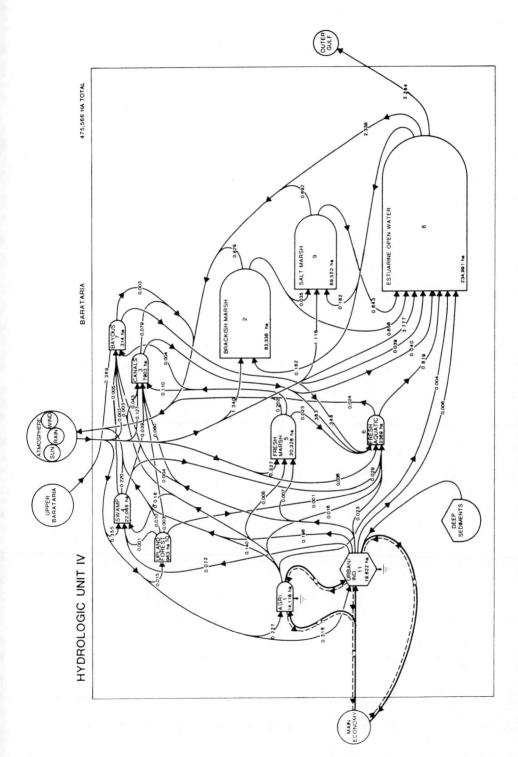

Figure 5-3. Conceptual model of the Barataria hydrologic unit (from Costanza et al., 1983).

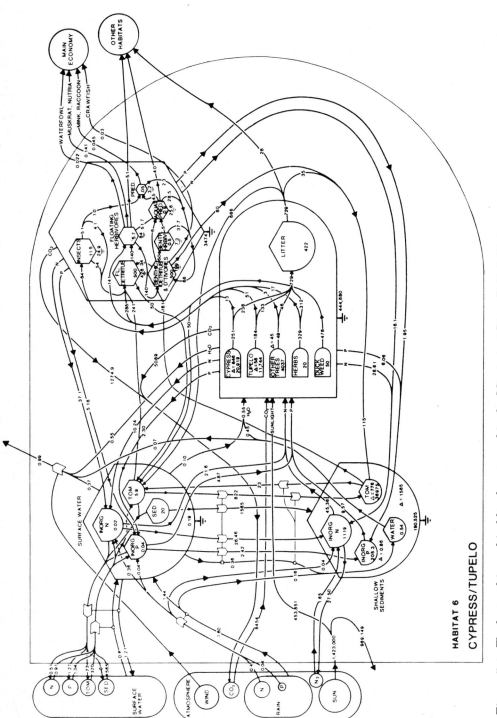

Figure 5-4. The lowest hierarchical level portrayed in the Mississippi River Deltaic Plain Region—conceptual model of the cypress–tupelo swamp (from Costanza et al., 1983).

Hydrologic Unit Level

1. *Role of Wetlands in Fishery Production*: Fishery production in the MDPR is believed to be dependent on organic matter produced in wetland habitats. Differences in wetland habitat composition among hydrologic units may be reflected in fishery harvest differences among hydrologic units. Harvest data, however, are lacking or of poor quality. The development of quantitative data on carbon flow through hydrologic units should be a major objective of future research.

2. *Hydrologic Modification*: Cultural changes (e.g., canal construction, spoil bank and levee construction, and impoundments) disrupt the hydrology that integrates coastal ecosystems. The cumulative effects of hydrologic alterations are most apparent at the hydrologic unit level.

3. *Water Quality*: Eutrophication and the introduction of toxic substances affect water quality throughout some hydrologic units. The optimal management of water quality requires knowledge about the fates and effects of nutrients and toxic substances.

4. *Salt Water Intrusion*: Hydrologic modifications and natural processes have allowed the landward progression of isohalines in many of the drainage basins, resulting in loss of habitat and municipal water supply problems.

Habitat Level

1. *Human-introduced Stresses*: Many cultural processes disturb specific habitats. A marsh area may be sublethally stressed by partial impoundment; a body of open water may be made eutrophic; or the soil in an agricultural habitat may be depleted of organic matter.

2. *Estimation of Resource Productivity and Value*: The economy of the MDPR benefits from and depends upon the products and services of various natural habitats. Better estimates of the rates and value of ecological production from each habitat are needed.

The U.S. Fish and Wildlife Service is constantly being asked to render opinions as to the effects of specific activities on the vitality of habitats and populations of organisms in coastal regions. In order to facilitate the decision making process and guide resource management and coastal planning, they have requested research teams around the U.S. to prepare ecological characterization studies. The studies describe the important components and processes of selected ecosystems and provide an understanding of their relationships by synthesizing and integrating existing physical, biological, and socioeconomic information. Only when humans understand how ecosystems function will they be able to effectively manage their natural resources and prudently guide developments generated by social and economic demands. The hierarchical, conceptual modelling approach exemplified by the MRDP study has proven to facilitate an understanding of the relationships operating at various levels of ecological organization by minimizing problems associated with differences in scale and duration between physical and biological events. The approach allows decision makers to look at cause and effect relations and to systematically evaluate the effects of specific activities.

Des Allemands Urban Runoff—Swamp Eutrophication Models

Hopkinson and Day (1980a,b) described two models that investigated relationships between urbanization on high lands, hydrology, and wetland eutrophication. In the first model the interaction between changing land use on uplands and storm water and nutrient runoff to adjacent coastal swamps was investigated. In the second model, Hopkinson and Day compared the effects of sheet flow versus channelized stream flow through swamp forests on rates of stormwater runoff from the uplands, on swamp productivity and nutrient dynamics, and on eutrophication of receiving water bodies.

The des Allemands swamp forest system is located in the headwaters of a large coastal basin bordered by two distributaries of the Mississippi River and the Gulf of Mexico. Natural levees along the distributaries direct runoff water into the interior wetlands and from the headwaters to the Gulf of Mexico. Although it is a nontidal system, water levels are controlled to some extent by lower estuary conditions. The objectives of the des Allemands models were:

1) To quantitatively predict present and future rates of nutrient and water runoff from the natural levee uplands as a result of changing land use during urbanization.

2) To ascertain the effects of spoil banks (artificial levees composed of dredged material from canal bottoms that is placed along canal banks) and drainage canals on water flow through and drainage in the swamp forest.

3) To evaluate the feasibility of routing upland runoff directly through backswamp areas rather than through drainage canals.

4) To determine the efficacy of sheet flow water runoff through backswamps in reducing the nutrient load to receiving bodies of water.

The United States Environmental Protection Agency sponsored the development of a comprehensive mathematical model capable of simulating urban storm water runoff and receiving effects for lakes, rivers, or wetlands. The Storm Water Management Model (SWMM) is divided into separate blocks. Hopkinson and Day used the RUNOFF block to simulate runoff from uplands and the RECEIVE block to simulate swamp and lake hydrology and nutrient dynamics. Figure 5-5 shows a conceptualization of the SWMM model using Odum symbols. Runoff from uplands was modeled using basic hydrologic equations. Real storm events are modelled on the basis of rainfall inputs and land use characteristics. Rainfall runs off the uplands, depending on the physical, chemical and biological conditions existing on the uplands. Runoff water transports available nutrients. Runoff is calculated from an accounting of rainfall, infiltration, evaporation, detention depth, overland flow, and canal flow. Infiltration is expressed by Horton's exponential function:

$$I_t = f_0 + (f_1 - f_0) e^{-at} \qquad\qquad (5\text{-}1)$$

where, I_t = infiltration, f_0 = minimum infiltration rate, f_1 = maximum infiltration rate, a = decay coefficient of infiltration.

A modified Manning's equation is used to control water runoff after a critical water depth has been reached:

$$V = (1.49/n) (D_2 - D_d)^{1/3} S^{1/2} \qquad\qquad (5\text{-}2)$$

where, V = velocity, n = Manning's roughness coefficient, D_d = detention depth, S = land slope, D_2 = amount of water remaining on surface following infiltration.

Flow in canals is calculated as:

$$V = (1.49/n) R^{2/3} S^{1/2} \qquad\qquad (5\text{-}3)$$

where, R = hydraulic radius.

The continuity equation is used repeatedly to account for water depths and volumes at all locations. Nutrients are washed off the land surface at a predetermined logarithmically decreasing rate after water runoff commences. Quantity and concentration of nutrients exported is determined from solution of mass balance equations.

To implement this portion of the model Hopkinson and Day identified all the upland drainage basins and then determined hydraulic and nutrient properties of each after accounting for changes in land use projected to occur between 1975 and 1995.

The RECEIVE block of SWMM is a finite element model. Hopkinson and Day represented the area to be modelled by a network of nodal points connected by channels. The points and channels are idealized hydraulic elements. Water movement in the network is calculated by simultaneously solving the equations of motion and the continuity equation to produce a temporal history of water height, velocity, and discharge. Only two dimensional flow is modelled; it is assumed that the water column is vertically homogeneous. Nutrient dynamics are modelled using predetermined decay coefficients for lake, canal, stream, and flowing and stagnant backswamp habitats. A minimal nutrient concentration value was established for each habitat to represent the sediment buffering effect on nutrient levels.

Several studies were conducted prior to the modelling work of Hopkinson and Day which indicated that there were fundamental ecological changes occurring in the swamp forest ecosystem as a result of urbanization and increased agricultural production on the surrounding uplands. It had been documented that lakes that were once clear and coffee-colored and that had abundant game fish populations had become characterized by frequent algal blooms, periodic fish kills, and "trash" fish populations. Craig and Day (1977) and Day et al. (1977) found strong correlations between the density of drainage

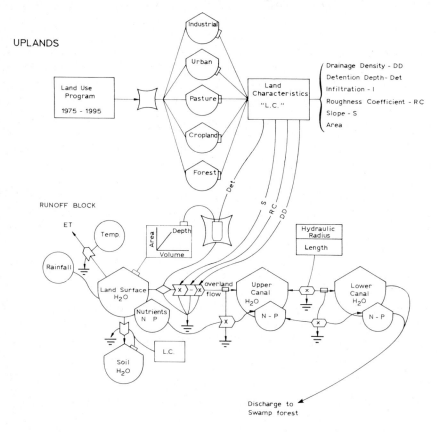

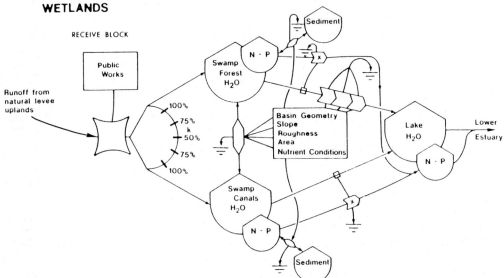

Figure 5-5. Conceptualization of the Storm Water Management Model (SWMM) of a Louisiana swamp showing a) uplands and b) wetlands (from Hopkinson and Day, 1980 a,b; reprinted by courtesy of Springer-Verlag, Inc.)

canals in the swamp, eutrophication of lake waterbodies, and decreasing productivity of impounded swamp forests. The Louisiana Office of State Planning predicted in 1975 that the uplands surrounding the swamp ecosystem would experience substantial development in the following 20 years reflecting the secondary growth associated with the construction of the Louisiana Offshore Oil Port terminal facility. The modelling study of Hopkinson and Day was a first attempt to integrate several studies that had been conducted in and around the swamp forest ecosystem.

Most values for equation parameters were obtained from the literature rather than measured in the field. Model validation was not intensely field verified; rather, rigorous sensitivity analyses identified those parameters that most critically affected model output. The authors found that the parameters which actually affected model results the most were those in which they had the greatest confidence. Hopkinson and Day stated that future scientific effort should be directed toward field validation of the models with special attention directed at parameters most critical to model results.

Simulation results showed that runoff volumes and nutrient loading to the swamp forest would increase greatly by 1995 if the projected changes in land use occurred. The model predicted that nutrient runoff to the swamp over the course of a year would increase 25 and 18 percent for nitrogen and phosphorus, respectively. The greatest increase in loading was noted for small rainfalls (<3.2 cm/day) when the projected changes were up to 210 percent greater. Sensitivity analyses showed that the increased areal extent of impervious areas was largely dictating the greater runoff volumes to the swamp. Simulations of swamp hydrodynamics and nutrient dynamics showed that spoil banks retarded water exchange between backswamps and streams and caused periods of prolonged ponding in the backswamps. It was found that discharge of upland runoff could be increased by removing all spoil banks and introducing it directly to backswamps rather than to drainage canals. If all barriers to overland flow were removed, a considerable portion of upland runoff would flow as sheet flow through the swamp rather than only through streams and canals. Although absolute runoff to the swamp was shown to increase under 1995 conditions, simulations showed that nutrient loading of the downstream lake in the des Allemands system could be decreased about 25 percent for both nitrogen and phosphorus.

Summers—McKellar North Inlet Estuarine Models

A series of reports (Summers et al., 1980; Summers and McKellar, 1981) describe a third generation large ecosystem model of carbon/energy flow in a marsh/estuarine system typical of southeastern U.S. Atlantic coastal salt marshes. Although energy is explicitly modelled, parameters are derived from data on carbon and a carbon to energy ratio of 1g:10 kcal. The model emphasizes intra- and inter-system flows of four major subsystems of the marsh/estuarine complex with adjacent oceanic waters. The model is basically a food chain model that uses the physical parameters, sunlight, temperature, sea level and tidal mixing, as the primary controls.

The modelling work reported is an evolving integral part of a major research program conducted by University of South Carolina scientists working at the Belle W. Baruch Foundation's coastal property. The completely natural area provides an unmodified natural system ideal for investigation of the factors controlling the exchange of materials between estuarine and coastal regions. The Baruch modelling work serves as an integral portion of the total research effort by summarizing and integrating field research and by elucidating profitable avenues of future research. Specific objectives of the third generation model are:

1) To estimate organic matter exchange with the nearshore continental shelf.

2) To elucidate the storages and sources of the offshore flow.

3) To determine the importance of tides and tidal mixing of estuarine water on controlling the exchange of matter between estuary and shelf.

4) To determine which model components and coefficients are important to system behavior.

5) To demonstrate the utility of systems theory techniques in general with suggestions for their application to large ecosystem projects.

Figure 5-6 shows the 19 compartments and the variety of controls used in the model. Spatial variation is accounted for by dividing the ecosystem into four subsystems: emergent marsh, oyster reef, benthos, and aquatic water column. Solar insolation and water temperature are the major external factors controlling component metabolism and trophic transfers. Photosynthesis is an unweighted multiplicative function of solar insolation, water temperature and producer biomass. The major shortcomings of this formulation are the absence of a nutrient control, the varying Q_{10} value (a doubling of rate occurs as temperature in the centigrade scale increases 10°C), and a growth-light intensity relationship that lacks an optimum intensity value and a Monod response. For each biotic compartment, the exponential effect of temperature on respiration is combined with a quadratic function of biomass to represent the crowding effect of individuals within the compartment. Trophic transfers are controlled by biomasses of both linear and donor compartments as well as by temperature.

Some of the greatest effort in this modelling series is in the elucidation of exchange between the marsh and the estuary and between the estuary and the nearshore region. Exchange from the marsh to the water is conceptualized as a function of the size of the donor compartment (e.g., the pool of particulate organic material on the marsh surface), the proportion of marsh covered at high tide, and the proportion of donor stock that can be outwashed if the marsh is 100 percent covered. Annual fluctuations in sea level dictate the percentage of marsh covered at high tide. Hopkinson and Day (1977) used a similar relationship to control the movement of detritus from a Louisiana salt marsh. The shortcoming of this approach is that flushing energy is not necessarily related to the monthly sea level. More important is the number of complete flushings per unit time, not the depth of water overlaying the marsh at high tide. If the marsh does not fully drain at low tide, there will be little flushing except of dissolved materials.

The exchange of material between the estuary and the sea is modelled as a function of the concentration gradient and a remixing coefficient. The remixing coefficient describes the portion of the ebb flowing water which re-enters the estuary on flood tide. Values used in the model vary from 10 to 90 percent and are related to the strength of the longshore current at different times of the year.

The type of controls in the Summers-McKellar model greatly contrasts the approach taken in the Wiegert-Wetzel and Wetzel-Christian models. The latter models use a great deal of detail in characterizing production and feeding of organisms. They use mathematical formulations with "biological reality" to describe resource and space limitation. In strong contrast, the Summers-McKellar model provides great detail in the inclusion of major driving forces basic to the functioning of all salt marsh/estuaries. Control of system behavior in their model lies in the input of seasonal variations in the driving forces. The approach taken by Summers-McKellar is similar to that taken by Hopkinson and Day (1977) in their model of the salt marsh ecosystem.

Although the salt marsh is a stable ecosystem in the sense of repetitious annual patterns of biomass for the major flora and fauna, Summers and McKellar felt that validation of their model could not be made by comparing output with the identical data used in the initial construction of the model. A good fit could be expected merely from the interdependence of the two data sets. Therefore, Summers and McKellar collected an additional independent data set to compare with their model output. Statistical tests of correspondence between the independent and model data sets were used to validate the model. Sensitivity analyses were conducted to reveal those pathways and mechanisms most important in controlling system behavior. One of the more interesting insights developed from sensitivity analysis was the importance of a *Spartina alterniflora*-marsh algae interaction that greatly affected system net productivity. As presently formulated, with increasing values of annual mean biomass of *Spartina*, annual system net production would increase with a corresponding decrease in algal biomass. With decreases in the biomass of *Spartina*, net system production would also increase due to increasing algal biomass and productivity.

As with most large ecosystem models, values for standing crops and equation parameters came from a combination of literature, field and mass balance sources. The annual trends of standing stocks reflected annual patterns measured in the field for the most part. Only one component (out of 19) failed the validation procedure. The authors conclude that because the major aspects of the model structure and outputs were not invalidated, the model presents a reasonable working hypothesis for the pathways of carbon flow in the system, between the marsh and the estuary, and between the estuary and oceanic waters.

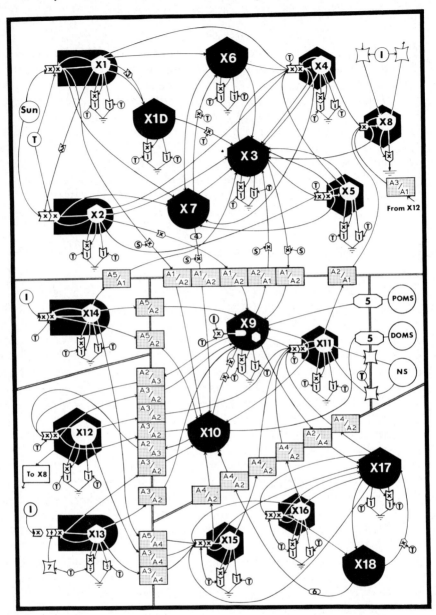

Figure 5-6. Conceptual model of energy flow in the North Inlet marsh–estuarine ecosystem. Components: sun or I, insolation; T, water temperature; S, sea level; POMS, particulate organic matter concentration offshore; DOMS, dissolved organic matter concentration offshore; NS, nekton density offshore; A1, surface area of marsh; A2, surface area of water column; A3, surface area of oyster reef subsystem; A4, surface area of subtidal benthos; A5, surface area of mudflats; 1, Q_{10} exponential relationship between temperature and respiration; 2, exponentially attenuated insolation; 3, time delay; 4, Fickian diffusion; 5, exchange between estuarine and coastal waters as affected by tidal remixing and intertidal volume; 6, Fickian diffusion between sediment and water column; 7, temperature switch for oyster reef macroalgal productivity (from Summers et al., 1980).

The output from the model illustrated several systems-level characteristics of the South Carolina marsh/estuarine system. The marsh is distinctly autotrophic, while the benthos, water column and oyster reefs are heterotrophic. The marsh subsidizes the remaining subsystems of the marsh/estuary to the extent that the whole system is autotrophic ($16gC/m^2 \cdot yr$) and has a net export of $19gC/m^2/yr$ to the sea.

Modelling is an integral ongoing component of the overall research effort by the Belle W. Baruch Foundation. It was alarming to the modellers that of 39 (out of 122) model parameters that had a moderate to high sensitivity, most were in the low data availability category. Identification of critical data needs was one of the objectives of the overall modelling effort.

Sapelo Island Salt Marsh Models

The salt marsh-estuarine complex adjacent to Sapelo Island, Georgia, U.S., has been the focus of an ecosystem-level simulation modelling effort since the early 1970s. A review of its history illustrates not only the philosophical and applied nature of the continuing studies but also the inherent evolutionary pattern of most large-scale modelling programs. The salt marsh modelling effort has served as both a simulation analysis tool (Wetzel and Wiegert, 1983) and a generator of testable hypotheses relative to the principal components, fluxes and controls on carbon dynamics in ecosystems dominated by *Spartina alterniflora* (Wiegert et al., 1975; Wiegert and Wetzel, 1979; Wiegert et al., 1981, 1984). These food chain-based models include much "biological reality" by their incorporation of population-level controls.

As with the University of South Carolina marsh research program, the University of Georgia program at Sapelo Island includes modelling as an integral part of the total research effort. Field research at Sapelo Island dates back to the early 1950s and the initial modelling effort was developed to summarize and integrate the field research and to elucidate profitable avenues of future research. Specific objectives of each modelling effort became more sophisticated as basic knowledge of ecosystem function improved.

The basic compartmental and flow structure of the salt marsh model employed (with several revisions in flow and/or mathematical structure) for the past eleven years is composed of 14 compartments (seven abiotic and seven biotic) that represent the principal components exchanging carbon between salt marsh-tidal creek sediments, water and the atmosphere (Figure 5-7). In realization of the importance of anaerobic sediment processes, the model specifically separates above- and below-ground biomass and production of *Spartina* and incorporates sediment detail with an anaerobic detrital community. Although the model is process-oriented rather than species-oriented, there is much more resolution at the population level in this modelling effort than in the Summers-McKellar salt marsh modelling work.

The mathematical structure of the model is relatively unique for ecosystem-level simulation models and is based on a proposed minimum set of "laws" governing the growth and interactions of populations (Wiegert, 1973, 1975). The functional forms of the flux equations generally fall into two categories. The first, non-linear feedback controlled equations given in the general form as:

$$F_{ij} = p_{ij} \cdot t_{ij} \cdot x_j \cdot f_{ij} \cdot f_{jj} \tag{5-4}$$

where,

F_{ij} = flux of carbon from compartment i to j

p_{ij} = preference of recipient j for resource i (0 > p > 1)

t_{ij} = maximum specific rate of transfer (g/g/day)

x_j = recipient compartment biomass

f_{ij} = negative feedback donor control due to scarcity of resource (0 > f_{ij} > 1)

f_{jj} = negative feedback recipient control due to space (0 > f_{jj} > 1)

were used for simulating abiotic-biotic and biotic-biotic exchanges. The feedback terms were functions of measurable parameters and given in the general form as:

$$f_{ij} = \{1.0 - [(a_{ij} - x_i)/(a_{ij} - g_{ij})]\} \tag{5-5}$$

where, a_{ij} = threshold response biomass below which the resource becomes limiting to uptake by recipient j, g_{ij} = resource biomass at and below which the resource is not available for uptake by recipient j, for the resource controlled negative feedback term, and as:

$$f_{jj} = 1.0 - \{[(1.0 - x_j - a_{jj})/(g_{jj} - a_{jj}) \, C_{ij}]\} \tag{5-6}$$

where, a_{jj} = threshold recipient biomass where space of some spatially related property limits the rate of growth of the j compartment, g_{jj} = the maximum allowable or maintainable biomass of compartment j, C_{ij} = metabolic correction for maintenance @ x_j = g_{jj} such that F_{ij} = sum of maintenance costs provided resource is not limiting, for the self or recipient controlled feedback term.

The second, linear donor controlled equations given in the general form as:

$$F_{ij} = r_{ij}x_i \tag{5-7}$$

where, r_{ij} = specific rate coeficient (g/g/day), were used for simulating abiotic-abiotic (e.g., physical exchanges) and biotic—abiotic exchanges (e.g., respiration, excretion, mortality).

Unlike the Summers-McKellar models, external or environmental forcing functions (e.g. light, temperature, sea level, tides) were not modelled explicitly,

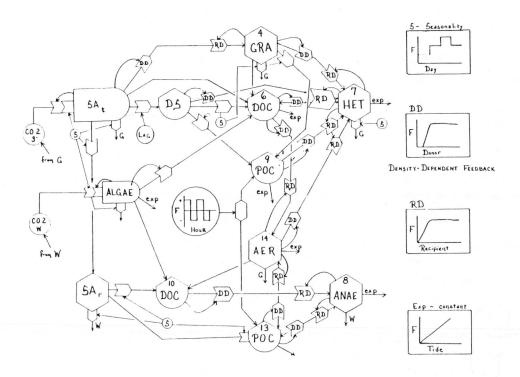

Figure 5-7. Conceptual model of compartmental and carbon flow structure in the Sapelo Island marsh-estuarine ecosystems (redrawn from Wiegert and Wetzel, 1979). SA_t = *Spartina alterniflora* shoots; SA_r = *Spartina alterniflora* roots; DS = dead standing grass; GRA = grazers; DOC = dissolved organic carbon; HET = all aerobic organisms in water; POC = particulate organic carbon; AER = all aerobic organisms in or on the sediment; ANAE = all anaerobic organisms in the sediment; DD = donor controlled, density dependent feedback function; RD = recipient controlled density dependent feedback function; G or W = respiratory loss of CO_2 to air or water; exp = tidal or bedload transport to sound-coefficient set so that a certain percentage was removed daily; S = seasonally varying coefficients.

rather, specific rate coefficients for biotic exchanges were varied seasonally. Following the initial version however, the model included algorithms for simulating carbon export due to tidal exchange that were internally driven.

A very interesting aspect of the mathematical structure is the use of the non-linear feedback controlled equations (Wiegert, 1973, 1975). To a large extent this aspect reflects the population-level background of many of the University of Georgia scientists working at Sapelo Island. The type of information required for this approach is not collected routinely in most salt marsh research programs and as such the approach is not widely used elsewhere. The mathematical derivation of the feedback functions represents a hypothetical statement regarding control. Because all parameters defining the two feedback functions are theoretically measurable, the controls are experimentally testable. By adopting this approach, the modelling effort and experimental research effort were closely coupled from the initiation of the program.

Over its eleven-year history, the salt marsh model evolved through six versions. Versions 1, 3, and 6 have been published, (Wiegert et al., 1975; Wiegert and Wetzel, 1979; Wiegert et al., 1981, 1984) while versions 2, 4 and 5 were transient and represented development-revision stages.

The first version was implemented to: 1) identify where information and data were missing relative to the conceptual model and aid in guiding the overall research effort; 2) identify parameters and/or controls that governed model behavior; and 3) qualitatively predict the overall behavior of the salt marsh ecosystem relative to its role as a potential source or sink of carbon for contiguous systems and which components were principally responsible for the observed behavior. The parameters associated with *Spartina*, algae, algal consumers and bacteria made up the majority of those meeting sensitivity criteria and were proposed as the areas where more and/or new research effort would best be allocated. Relative to the salt marsh carbon source or sink question, the model answer was unequivocal: the salt marsh functioned as a source of organic carbon.

Three years intervened between the publication of the first and third versions of the model. During this period, research was focused on refining the input data relative to the sensitive parameters, revising in part the flow and feedback structure to reflect new data and/or information and incorporating flows in the model to simulate tidal exchange.

Wiegert and Wetzel (1979) summarized the results of version 3. Two series of simulation studies were undertaken to corroborate the model (sensu Caswell, 1976). For the first series, a simple algorithm employed for tidal export was varied over a range of 0 to 50 percent (i.e. fractions of specific compartments exported per day) and the results compared to annual average compartment standing stocks and net carbon balance for the marsh ecosystem. The importance of tidal export relative to carbon balance for the marsh system was clearly indicated. Although the revised model could not address the mechanisms or fates of the exported material, the results confirmed the original conclusion that the marsh was a carbon source and suggested that the mechanisms governing exchange should be more thoroughly evaluated. In the second series of simulation studies with version 3, Wiegert and Wetzel (1979) reported the results of comparing model predictions with four field experiments. The results were impressive in that for all comparisons, model predictions and experimental (field) results were in agreement. This added confidence in employing the model as an analysis- and hypothesis-generating tool.

The sixth and final version of the model was directed toward increasing the resolution of the mechanisms governing tidal export, as well as revising these parameter values for which new data were available. New values for parameters controlling tidal export were available from a study of the hydrology of the Duplin River, the main tidal channel draining the marshes adjacent to Sapelo Island (Imberger et al., 1983). In addition to changing the methods of calculating tidal exchange based on hydrodynamic constraints, model simulations also evaluated the effects of storms and certain changes in biologically-mediated pathways. The result of these revisions was to lower the predicted export of carbon by about 50 percent and to predict an annual accumulation on the marsh surface. A small amount of organic carbon was in excess and could not be accounted for.

Based on the model simulation results, the authors proposed three alternate hypotheses that could account for the excess: 1) bedload transport, 2) storm frequency greater than that simulated, and 3) biological vectors of movement. Version 6 represents the latest simulation analyses reported for this ecosystem modelling effort, although new efforts have already been undertaken (R.G. Wiegert, personal communication). This effort, over its eleven-year history, illustrates the variety of roles a model can and often does play in both the organization and analysis of ecosystem-level research. The models have provided insight into the ecology of coastal wetlands that either would not have been realized or at least would have been delayed in time. During the course of development of the model versions, various submodels of specific processes or different compartmentalization schemes were constructed and simulated. We summarize later the results of one of these on simulation model studies of the interactions between detritus derived from *Spartina*, microbes and microbial grazers that was a direct outgrowth of the salt marsh modelling effort. Wiegert et al. (1984) report on the results of a second submodelling effort directed at detailed studies of carbon exchange via macroheterotrophy. New generations of the salt marsh model will include information and material flows for both carbon and nitrogen but to accomplish this a second generation of simulation models will be needed with complete revision of compartmental and flow structures.

A Microbially-Linked Model of Carbon Flow in Estuarine Waterbodies

Christian and Wetzel (1978) and Wetzel and Christian (1984) described a series of two models designed to represent the principal components and interactions between detritus derived from *Spartina alterniflora* microbes and consumers in salt marsh waterbodies. The initial work was an outgrowth of the Wiegert-Wetzel et al. salt marsh modelling program discussed previously. There was within the salt marsh program considerable expertise and experimental work on marine aquatic food chains. Most of this information was either excluded or lumped almost beyond recognition within the ecosystem-level marsh modelling effort. The impetus for the microbial models was provided in part by the desire to develop a more realistic systems analysis tool for microbial interactions than the salt marsh model was capable of and to investigate via model simulation various alternate hypotheses regarding the coupling between microbes and higher-level consumers; i.e. the classical paradigm that the high secondary productivity of estuaries is directly linked to vascular plant detritus input and microbial degradation. The second version of the microbial model is a carbon-based food chain model containing nine compartments, four of which are biotic (Figure 5-8). As with the Wiegert-Wetzel et al. salt marsh models, there is emphasis on biologic reality in the form of feeding threshold and space limitation controls but little inclusion of the physical forcing functions that are unique to the estuarine environment. As with the salt marsh model, abiotic-biotic and biotic-biotic exchange pathways were modelled with non-linear, donor and/or recipient feedback controlled functions and abiotic-abiotic and biotic-abiotic pathways were modelled with linear, donor controlled functions.

The first version of the model was general and strictly a theoretical exercise intended to: 1) summarize the existing literature, 2) determine which parameters governed model behavior and 3) propose, based on simulation studies, the conditions necessary for stability of the conceptual system. Indirectly, the model examined the energetic feasibility of a direct micro-macroconsumer trophic link.

Based on their first simulation studies, Christian and Wetzel (1978) concluded that optimum stability for the particle-microbe complex would be characterized by a microbial component that: 1) has a high preference for dissolved organic substrates, 2) has short generation times, 3) has greater than a 50 percent growth yield, and, surprisingly, 4) is not heavily grazed. They also suggested that a significant portion of the microbial population should be free-living and based on thermodynamic constraints it is unlikely that a macroheterotrophic population could rely solely on particles with attached bacteria as an energy or nutrient source.

Both the compartmental and mathematical complexity were increased in the second version of the microbial food chain model. The major revisions in model structure were dividing the original microbial compartment into attached and free-living components, partitioning dissolved organic carbon substrates into fast (labile) and slow (refractory) components with different inputs, and including tidally-driven sediment exchanges with the water column, detritus, particle-attached, microbe compartment. Model simulation experiments were directed toward evaluating the potential for carbon resources and heterotrophic grazers controlling predicted dynamics of the microbial compartments. Specifically, Wetzel and Christian (1984) reported the results of simulations relative to the effects of 1) carbon resource allocation by particle-attached bacteria, 2) sensitivity of the microbial compartments to various dissolved organic carbon inputs, 3) potential for grazing control on microbial dynamics by an assumed macroheterotrophic consumer, and 4) potential for grazing control on microbial dynamics by an assumed microheterotrophic consumer.

The authors concluded that they were able to predict reasonably well the behavior of particulate and dissolved organic carbon, free-living bacteria and microconsumers when compared with actual field data and the literature. However, the model, given reasonable constraints on the parameter values and initial conditions used for simulation, did not adequately simulate the dynamics of particle-attached bacteria; i.e. the model consistently predicted a high density of bacteria on particles regardless of inputs or grazing rates reported for microconsumers (microflagellates in the model). From these results alternate hypotheses were suggested that could account for the predicted behavior. Interestingly, the second and more complex version indicated that a direct microbe-macroconsumer trophic link in the estuarine water column could not be supported. If parameter values for growth and metabolism typical of those reported for macroheterotrophs in the water column (e.g., copepods) were input as data to the model, grazing pathways exercised no control over microbial dynamics, and as importantly, the macroconsumer compartment did not survive over time. Only by inputting data and initial conditions typical of microheterotrophic populations were the predicted dynamics of their model corroborated by the results and conclusion of other experimental or field investigations. They also concluded, based on specific sensitivity analyses of the microheterotrophic compartment, that the actual parameter values (i.e. resource preferences, maximum feeding rates, assimilation efficiencies, and rates of respiration) governing growth of the populations were very sensitive (sensu Wiegert et al., 1975) and therefore should be more thoroughly evaluated relative to growth and dynamics of the population per se, and as potentially important control on both energy flow and nutrient cycles. As we discuss later, these conclusions may stem in some part from the inclusion of population-level detail in model control at the expense of outside forcing function control.

This modelling effort is a good example of the benefits of a strong coupling between field experimentation and modelling. Based on the alternate hypotheses proposed by this modelling effort to explain the behavior of attached-particle microbes, new research has been proposed to test these results using a combination of model simulation analyses, field studies and laboratory microcosm experiments.

North River Ecosystem Model

In a collaborative field and modelling effort, the University of New Hampshire Complex Systems Research Center and the Marine Biological Laboratory

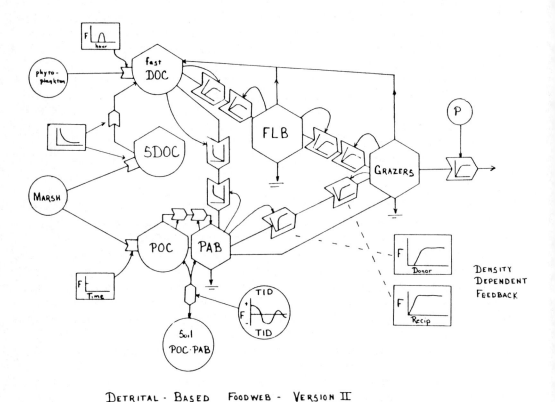

DETRITAL - BASED FOODWEB - VERSION II

Figure 5-8. Conceptual model of compartmental and flow structure used to simulate the interactions and dynamics of carbon in a detrital-based food web in the water column of Sapelo Island, Georgia estuary. Fast DOC = labile dissolved organic carbon; SDOC = recalcitrant dissolved organic carbon; POC = particulate organic carbon; FLB = free-living bacteria; PAB = particle attached bacteria; P= predation; F= flux of carbon.

Ecosystems Center investigated nutrient dynamics in a tidal, freshwater marsh and river in coastal Massachusetts, USA. Attention was focused on the processing of nitrogen as water moved on and off fringing river marshes and flowed through the system. A hydrodynamic model was developed to integrate field and laboratory data. The one-dimensional water flow model had the capability to account for material transport and to incorporate material decay coefficients. As such, the model could describe the behavior of both conservative and nonconservative constituents.

Modelling was an integral part of the North River research program from the outset. Initially, conceptual box and arrow models were used to organize and synthesize literature information and to plan a coherent field and laboratory research plan. It was only after a period of ambitious field work that a need arose for development of a simulation model. The objectives of the model were:

1) To calculate material budgets for the North River system.

2) To determine the nature and magnitude of resident biotic transformations of nitrogen.

3) To predict the spatial and temporal distribution of constituents throughout the estuary.

4) To determine the influence of water movement alone on the distribution of nitrogen within the river stretch.

5) To evaluate the relative importance of biotic uptake (benthic, phytoplankton and marsh) on controlling the distribution of nitrogen within the stretch of river.

Vorosmarty et al. (1981) adapted a detailed one-dimensional tide-propagation model developed by Lee (1971). The model is described as a finite element model where elevation, area and bottom friction coefficients are specified for each element. The model has the capability of accomodating forcing functions at system boundaries. For the North River simulation, forcing functions included tidal inputs at the downstream end member and freshwater inputs at the upstream end members. The model explicitly solves the equation to determine tide height and water flow at each element. The North River basin was geometrically conceptualized (Figure 5-9) as a "Y-shaped" assemblage of 18 discrete river-marsh blocks (elements). Each block represented a well-mixed reservoir of both marsh and river components. Marsh components represented only storage volumes which flanked central river channels. Advective flux was confined to river channels. Conservative material behavior was accomodated by a mass balance approach where water flow was multiplied by material concentration. Addition of terms which altered material balances at each block permitted nonconservative material behavior to be modelled. Biotic controls on nitrogen dynamics in the North River included benthic remineralization and exchange across the sediment-water interface, phytoplankton uptake and exchanges between the water column and the marsh litter. The exchange between the water column and marsh litter was modelled as a zero-order process. Phytoplankton assimilation of nitrogen was modelled as a first-order recipient controlled process.

The North River model was calibrated and assumed validated when the simulated distribution of salinity closely matched field observations over an entire tidal cycle. The research team had observed, prior to modelling, a consistent patten of decreasing inorganic nitrogen concentration in a downstream direction in the North River. The calibrated hydrodynamics model was used with forcing function inputs of nutrients to predict that the great majority of the concentration decrease could be accounted for by water movement (dilution) alone. The model indicated, after incorporating biological transformation, that benthic and marsh uptake was responsible for 66 percent and 33 percent of total uptake, respectively. Close agreement between observed nutrient concentrations and simulated output indicated the model was able to predict both the direction and overall magnitude of the biotic uptake. The authors suggested that discrepancies could be due to hydrodynamic assumptions and the application of uniform zero-order nutrient exchange rates in an environment that is clearly heterogeneous. Sensitivity analyses were never conducted.

The model proved to be a useful tool for estimating material budgets for the North River. The calibrated model consequently precluded the necessity for the extremely time consuming *in situ* sampling techniques. Not only did the model provide important information on mass flux of nitrogen within the North River system, but it also provided interesting information on the relative importance of dilution and biotic processing on nitrogen dynamics and it predicted the spatial and temporal distribution of constituents throughout the river. The major shortcomings of the model were the inability to simulate the effect of natural levee diking on the marsh, the lack of biotic detail, and the lumping of marsh and river subsystems.

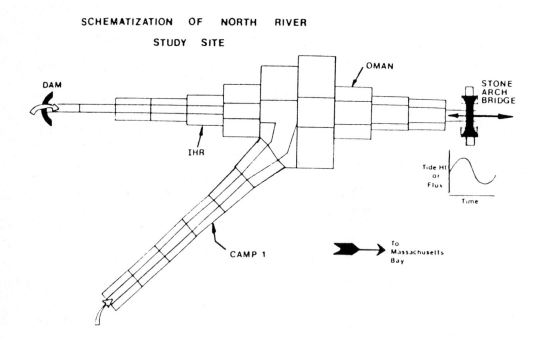

Figure 5-9. Integrated geometry used to model the hydrodynamics of the North River ecosystem (from Vorosmarty et al., 1983). Each rectangle represents a well mixed reservoir with marsh components represented as storage volumes flanking central channels.

It is unfortunate that a lack of continuing research funding caused the research effort to be terminated shortly after model construction. Considerable field data had been collected dealing with nitrogen transformations within marsh sediments, nitrogen nutrition of marsh macrophytes, and detrital decomposition on the marsh. Such data were never incorporated into a simulation model. The modified Lee model has strong similarities to the Storm Water Management Model that Hopkinson and Day (1980a,b) used to simulate water transport and nonconservative element behavior in the des Allemands swamp. The higher level of sophistication of the SWMM model enabled Hopkinson and Day to easily include much biotic detail and simulate the diking effect of creek bank levees on water ponding within the swamp forest. Despite the limitations of the Vorosmarty model, it seemed to give a reasonable simulation of water flux and behavior of nonconservative elements in the North River ecosystem. It proved useful for calculating a nitrogen budget of the system and was a helpful device for generating research hypotheses.

Interpretation of Sensitivity Analyses—A Note of Caution

Sensitivity analysis of simulation models provides one means of assessing the adequacy of component and parameter quantification. If sensitivity analysis of a selected parameter, the quantification of which is dubious or not available, indicates that model response is less than moderately affected, there is little justification to emphasize research efforts toward increasing precision in the estimation of that parameter. Primary areas for emphasized research would be directed where parameter sensitivities were high.

Another, perhaps more important, aspect of sensitivity analysis is its usefulness in revealing those pathways and mechanisms which most affect system behavior (Rosen, 1971). Intersystem comparison of controlling mechanisms could provide a means to characterize basic properties of various ecosystems and thus serve as a useful tool in the emerging field of comparative ecology. However, the usefulness of sensitivity analysis as a means to reveal factors controlling certain ecosystem behaviors is limited by the initial abstraction or conceptualization of the ecosystem structure (sensu Wiegert, 1979).

This can be very clearly demonstrated by comparing the differences in interpretations given to sensitivity analyses of the salt marsh models (Wiegert et al., 1975; Wiegert and Wetzel 1979; Summers et al., 1980; Summers and McKellar 1981). Wiegert and Wetzel concluded that the population-level parameters dealing with feeding thresholds, carrying capacities and space limitations were the primary controlling factors in the Sapelo Island marsh/estuarine ecosystem. In contrast, Summers and McKellar discovered that it was a combination of the "major driving forces basic to the functioning of the salt marsh ecosystem" that were the primary controlling factors. The major driving forces included sunlight, temperature, tidal action, sea level, and other forces external to the system.

As both models are of similar salt marsh systems that are typical of the sediment-rich, moderate tidal, low wave energy southeast U.S. coastal zone, the disparity between the conclusions concerning controlling factors can only stem from differences in model structure. As we mentioned previously, Wiegert and Wetzel (1979) included an abundance of population-level detail and "biological reality" in their mathematical representation of abiotic-biotic and biotic-biotic exchange pathways. To a degree this represented the background of the modellers and the preponderance of data on organism energetics. Tides and sea level, forces unique to coastal systems, were excluded explicitly as factors potentially controlling exchange. Tides were incorporated only as decay functions to remove preselected quantities for certain compartments. Sunlight and nutrient controls on primary productivity were not explicitly incorporated either. Summers and McKellar's mathematical expressions of exchanges were almost the mirror images of the Wiegert and Wetzel formulations. Whereas they emphasized the potential controlling role of external forcing functions, they almost completely excluded any provisions for space limitation, feeding thresholds or population carrying capacities.

Future examinations of the factors controlling certain system behaviors might profitably use a hybrid conceptualization of the salt marsh ecosystem and incorporate both the biological reality of the Wiegert et al. models and the physical forcing function reality of the Summers et al. models. A hybridization would not only allow the relative importance of the different levels of controlling factors to be ascertained, but would also be a step toward the formulation of a more generalized model which could·then be "fine tuned" to "fit" all salt marsh ecosystems. These models would be considerably more useful not only for research purposes but also for directing management efforts in coastal regions.

Conclusions

Our review of marsh/estuarine models demonstrates that ecosystem-level simulation and conceptual modelling have become useful tools in the arsenal of resource managers and research scientists. Modelling is the best technique available for organizing, integrating and synthesizing the voluminous amount of information that is collected in ecosystem studies. For resource managers, there is presently no better way available which allows rapid presentation of ecological information, easily permits examination of cause and effect relations, and facilitates systematic evaluation of the effects of specific activities. Models also serve as planning instruments and guides to research programs. As estuarine models continue to improve by better integrating basic principles controlling these physically-dominated ecosystems and as "a model" is developed which can contain the essence of all marshes, they will become more useful to coastal managers and research scientists.

Acknowledgment

This is a joint contribution from the University of Georgia Marine Institute, (Contribution No. 582), the Virginia Institute of Marine Science (Contribution No. 1367), and the Coastal Ecology Institute, Louisiana State University (Contribution No. LSU-CEI-87-02). This work was supported by the Georgia Sea Grant College Program, part of the National Oceanic Atmospheric Administration, U.S. Department of Commerce.

References

Bahr, L., R. Costanza, J. Day, S. Bayley, C. Neill, S. Leibowitz and J. Fruci. 1983. *Ecological characterization of the Mississippi deltaic plain region: a narrative with management recommendations.* U.S. Fish and Wildlife Service Report FWS/OBS-82/69. Division of Biological Services, Washington, D.C., 189 pages.

Caswell, H. 1976. The validation problem. Pages 313-325 in B.C. Patten, editor. *Systems Analysis and Simulation in Ecology, Vol. 4.* Academic Press, New York.

Christian, R.R. and R.L. Wetzel. 1978. Interaction between substrates, microbes, and consumers of Spartina detritus in estuaries. Pages 93-113 in M.L. Wiley, editor. *Estuarine Interactions.* Academic Press, New York.

Costanza, R., L. Bahr, C. Neill, S. Leibowitz and J. Fruci. 1983. The Mississippi deltaic plain region study: an application of ecological models to the analysis and management of a complex coastal region. Pages 669-671 in W. Lauenroth, G. Skogerboe and M. Flug, editors. *Analysis of Ecological Systems: State-of-the-Art in Ecological Modelling.* Elsevier, Amsterdam.

Craig, N.J. and J.W. Day, eds. 1977. *Cumulative impact studies in the Louisiana coastal zone: eutrophication and land loss.* Final report to Louisiana State Planning Office, Baton Rouge, 85 pages.

Day, J.W., T. Butler and W. Conner. 1977. Productivity and nutrient export studies in a cypress swamp and lake system in Louisiana. Pages 255-269 in M.L. Wiley, editor. *Estuarine Processes, Vol. II.* Academic Press, New York.

Fruci, J., R. Costanza and S. Leibowitz. 1983. Quantifying the interdependence between material and energy flows in ecosystems. Pages 241-252 in W. Lauenroth, G. Skogerboe and M. Flug, editors. *Analysis of Ecological Systems: State-of-the-Art in Ecological Modelling.* Elsevier, Amsterdam.

Gagliano, S., K. Meyer-Arendt, and K. Wicker. 1981. Land loss in the Mississippi River deltaic plan. *Trans. Gulf Coast Assoc. Geol. Soc.* 31:295-299.

Hopkinson, C.S. and J.W. Day. 1977. A model of the Barataria Bay salt marsh ecosystem. Pages 236-265 in C. Hall and J. Day, editors. *Ecosystem Modelling in Theory and Practice: An Introduction with Case Histories.* John Wiley, New York.

Hopkinson, C.S. and J.W. Day. 1980a. The relationship between development and storm water and nutrient runoff. *Environmental Management* 4:315-324.

Hopkinson, C.S. and J.W. Day. 1980b. Modeling hydrology and eutrophication in a Louisiana swamp forest ecosystem. *Environmental Management* 4:325-335.

Imberger, J., T. Berman, R.R. Christian, E.B. Sherr, D.E. Whitney, L.R. Pomeroy, R.G. Wiegert and W.J. Wiebe. 1983. The influence of water motion on the distribution and transport of materials in a salt marsh estuary. *Limnol. Oceanogr.* 28:201-214.

Lee, C.H. 1971. One-dimensional, real time model for estuarine water quality prediction. Ph.D. Dissertation, Massachusetts Institute of Technology, Cambridge.

Leibowitz, S. and R. Costanza. 1983. A preliminary input-output model of salt marshes in the Mississippi deltaic plain region. Pages 771-780 in W. Lauenroth, G. Skogerboe and M. Flug, editors. *Analysis of Ecological Systems: State-of-the-Art in Ecological Modelling.* Elsevier, Amsterdam.

Rosen, R. 1971. *Dynamical System Theory in Biology.* Wiley-Interscience, New York, 340 pages.

Summers, J.K., W. Kitchens, H. McKellar and R. Dame. 1980. A simulation model of estuarine subsystem coupling and carbon exchange with the sea. II. North Inlet model structure, output and validation. *Ecol. Modelling* 11:101-140.

Summers, J.K .and H. McKellar. 1981. A sensitivity analysis of an ecosystem model of estuarine carbon flow. *Ecol. Modelling* 13:283-301.

Vorosmarty, C., B. Moore, W. Bowden, J. Hobbie, B. Peterson and J. Morris. 1983. The transport and processing of nitrogen in a tidal, freshwater marsh and river ecosystem: modeling the roles of water movement and biotic activity in determining water quality. Pages 689-698 in W. Lauenroth, G. Skogerboe and M. Flug, editors. *Analysis of Ecological Systems: State-of-the-Art in Ecological Modelling.* Elsevier, Amsterdam.

Wetzel, R.L. and R.R. Christian. 1984. Simulation model studies on the interactions between carbon substrates, bacteria and consumers in a salt marsh estuary: Duplin River, Sapelo Island, Georgia, U.S.A. *Bull. Mar. Science* 35:601-614.

Wetzel, R.L. and R.G. Wiegert. 1983. Ecosystem simulation models: tools for the investigation and analysis of nitrogen dynamics in coastal and marine ecosystems. Pages 869-892 in E.J. Carpenter and D.G. Capone, editors. *Nitrogen in the Marine Environment.* Academic Press, New York.

Wiegert, R.G. 1973. A general ecological model and its use in simulating algal-fly energetics in a thermal spring community. Pages 85-102 in P.W. Geier, L.R. Clark, D.J. Anderson and H.A. Nix, editors. *Insects: Studies in Population Management, Vol. I.* Occasional Papers, Ecol. Soc. of Australia.

Wiegert, R.G. 1975. Simulation modeling of the algal-fly components of a thermal ecosystem: effects of spatial heterogeneity, time delays and model condensation. Pages 157-181 in B.C. Patten, editor. *Systems Analysis and Simulation in Ecology, Vol. 2.* Academic Press, New York.

Wiegert, R.G. 1979. Modeling coastal, estuarine, and marsh ecosystems: state-of-the-art. Pages 319-341 in G. Patil and M. Rosenzweig, editors. *Contemporary Quantitative Ecology and Related Ecometrics.* International Cooperative Publishing House, Fairland, Maryland.

Wiegert, R.G. and R.L. Wetzel. 1979. Simulation experiments with a 14-compartment salt marsh model. Pages 7-39 in R.F. Dame, editor. *Marsh-Estuarine Systems Simulation.* Univ. South Carolina Press, Columbia.

Wiegert, R.G., R.R. Christian, J.L. Gallagher, J.R. Hall, R.D.H. Jones, and R.L. Wetzel. 1975. A preliminary ecosystem model of coastal Georgia *Spartina* marsh. Pages 583-601 in L.E. Cronin, editor. *Estuarine Research, Vol. I.* Academic Press, New York.

Wiegert, R.G., R.R. Christian, and R.L. Wetzel. 1981. A model view of the marsh. Pages 183-218 in L.R. Pomeroy and R. G. Wiegert, editors. *The Ecology of a Salt Marsh.* Springer-Verlag, New York.

Wiegert, R.G., R.L. Wetzel and E.F. Vetter. 1984. *Aerobe: An expanded ecosystem model of carbon transformations and transport in a Georgia salt marsh.* 1984 ISEM Meeting, North American Chapter, Colorado State University, Fort Collins, Colorado, 5-9 August 1984 (Abstract).

6/ A DYNAMIC SPATIAL SIMULATION MODEL OF LAND LOSS AND MARSH SUCCESSION IN COASTAL LOUISIANA

Robert Costanza
Fred H. Sklar
Mary L. White
John W. Day, Jr.

A spatial simulation model was constructed to help understand the historical changes in the Atchafalaya/Terrebonne marsh/estuarine complex in south Louisiana and to project impacts of proposed human modifications. The model consists of 2,479 interconnected "cells," each representing 1 square km. Each cell in the model contains a dynamic, nonlinear, simulation model that has evolved out of the modelling work mentioned in the previous chapter. Variables include water volume and flow, relative elevation, sediment, nutrient, and salt concentrations, organic standing crop, and productivity. The model produces weekly maps of all the state variables and habitat types. Habitat succession occurs in a cell in the model when physical conditions change sufficiently, so that the new conditions better match the "signature" of another habitat. In this chapter we: 1) summarize the history of the Louisiana coastal land loss problem and suggested solutions; 2) briefly discuss the model's structure, data base, and degree of fit with historical data; 3) discuss the uses and implications of the model, particularly as regards the estimation of the impacts of canals and levees on coastal marsh systems; and 4) outline the potential interface between the model and management agencies to provide solutions to pressing coastal management problems.

The Land Loss Problem in Coastal Louisiana

Wetland loss in coastal Louisiana is a cumulative impact, the consequence of many impacts both natural and artificial. Natural losses are caused by subsidence, decay of abandoned river deltas, waves, and storms. Artificial losses result from flood control practices, impoundment, dredging, and subsequent erosion of artificial channels. Wetland losses also occur because of spoil disposal upon wetlands and land reclamation projects (Craig et al., 1979). Land loss has been defined "as the substantial removal of land from its ecologic role under natural conditions" (Craig et al., 1979).

Losses occur in three basic ways: (1) wetlands become open water due to natural or artificial processes (loss of this type may be caused by erosion, subsidence, or dredging to form canals, harbors, etc.); (2) wetlands are covered by fill material and altered to terrestrial habitat; and (3) wetlands can be wholly or partly isolated by spoil banks (Craig et al., 1979).

Wetland loss in abandoned river deltas was once compensated for by land building in the region of the active delta. Today, due to human intervention, there is a net loss of wetlands of 102 km^2 (39.4 mi^2) annually in coastal Louisiana (Gagliano et al., 1981).

Natural Wetland Loss

The deltaic plain is an area of dynamic geomorphic change. For the past several thousand years, the Mississippi River has followed a pattern of extending a delta seaward into the gulf in one area, and, after a few hundred years, abandoning it gradually in favor of a shorter adjacent route of steeper gradient (Morgan and Larimore, 1957). When a delta lobe is abandoned, active land building via sedimentation ceases and net loss of land occurs because of erosion and subsidence. Because of levee construction, the Mississippi River has been effectively "walled in," and presently most of the sediments and nutrients of the river are deposited in the deep Gulf of Mexico and are unable to contribute to the buildup or maintenance of the coastal wetlands (Craig et al., 1979).

The most important processes affected by lack of sediment input are the rates of sedimentation and net marsh accretion of both streamside and inland marsh types (Cleveland et al., 1981). DeLaune et al. (1978) found that marsh sites closer to natural streams were accreting at a higher rate than inland marsh sites (1.35 cm/yr vs 0.75 cm/yr), and only streamside marsh areas were accreting fast enough to offset the effects of subsidence. Similar patterns of accretion were noted by Baumann (1980), who observed a mean aggradation deficit of 0.18 cm/yr for 80 percent of the marsh in Barataria basin. He proposed that this mechanism was responsible for a large portion of the marsh currently being lost in the Barataria basin. Most of the sediments which allow streamside marsh to maintain its elevation come from resuspended bay bottom sediments (Baumann et al., 1984).

Barrier islands along the coast are a strong defense against marine processes and hurricanes. The tidal passes between the islands act as the control valves of the estuaries by regulating the amount of high salinity water, storm energy, etc., that enter the estuaries (Gagliano, 1973). The barrier islands of the Barataria basin are currently eroding; Grand Isle and Grand Terre are listed as areas of "critical erosion" by state and federal agencies. Limited coastal sand supply in Louisiana has caused one of the most serious barrier island problems in the country. Barrier island retreat rates are as high as 50 m/yr and loss rates of 65 ha (160 acres) per year have been reported (Mendelssohn, 1982).

Human-Induced Wetland Loss

Primary human activities that contribute to wetland loss are flood control, canals, spoil banks, land reclamation, and highway construction. There is increasing evidence that canals and levees are a leading factor in wetland loss (Craig et al., 1979, Scaife et al., 1983, Cleveland et al., 1981, Deegan et al., 1984). For example, in the period between 1962 and 1974, 18,138 ha of wetland in Barataria basin were drained or converted to water, with agricultural impoundments and oil access canals accounting for the largest acreages (Adams et al., 1976).

Mississippi River Levees

The leveeing of the Mississippi and Atchafalaya Rivers, along with the damming of distributaries, has virtually eliminated riverine sediment input to most coastal marshes. This has broken the deltaic cycle and greatly accelerated land loss. Only in the area of the Atchafalaya delta are sediment-laden waters flowing into wetland areas and land gain occuring (Baumann and Adams, 1981).

Canals

Canals interlace the wetlands of the Louisiana coastal zone. Natural channels are generally not deep enough for the needs of oil recovery, navigation, pipelines, and drainage, so a vast network of canals has been built to accommodate these needs. The construction of canals leads to direct loss of marsh by dredging and spoil deposition and indirect loss by changing hydrology, sedimentation, and productivity. Canals lead to more rapid salinity intrusion, causing the death of freshwater vegetation (Van Sickle et al., 1976). Canal spoil banks severely limit water exchange with wetlands, thereby decreasing deposition of suspended sediments. The ratio of canal to spoil area has been estimated to be 1 to 2.5 (Craig et al., 1979), indicating the magnitude of spoil deposition to wetland loss.

It has been estimated that between 40-90% of the total land loss in coastal Louisiana can be attributed to canal construction, including canal/spoil area and cumulative losses (Craig et al., 1979, Scaife et al., 1983). In the Deltaic Plain of Louisiana, canals and spoil banks are currently 8% of the marsh area compared to 2% in 1955; there was an increase of 14,552 ha of canals between 1955 and 1978 (Scaife et al., 1983). Barataria basin had a 0.93%/year direct loss of marsh due to canals for the period of 1955-1978 (Scaife et al., 1983). Canals indirectly influence land loss rates by changing the hydrologic pattern of a marsh, such as by blockage of sheet flow, which in turn lessens marsh productivity, quality, and the rate of accretion. Spoil banks specifically block the import of resuspended sediments which are important in maintaining marsh elevation in wetlands distant from sediment sources. In time, canals widen because of wave action and altered hydrologic patterns, and apparently the larger the canal, the faster it widens. Annual increases in canal width of 2 to 14% in Barataria basin have been documented, indicating doubling rates of 5 to 60 years (Craig et al., 1979).

Generally, where canal density is high, land losses are high, and where land losses are low, canal densities are low. The direct impacts of canals are readily measurable. For example, from 1955 to 1978, canal surface area accounted for 10% of direct land loss. The indirect influence of canals extends far beyond this direct loss. Craig et al. (1979) estimated the total direct plus indirect loss of wetland due to canals in 3-4 times the initial canal area alone. Although total canal surface area alone may not be a dominant factor in wetland loss, direct and indirect impacts of canals may account for some 65% or more of the total wetlands loss between 1975 and 1978 (Scaife et al., 1983).

Cumulative Impact of Canals

The canals, when viewed on a regional basin level, become a network ultimately resulting in higher rates of wetland loss (Craig et al., 1979), increased saltwater intrusion (Van Sickle et al., 1976), changes in the hydrology of the wetland system (Hopkinson and Day 1979, 1980a, 1980b), a reduction in capacity for wetlands to buffer impacts of large additions of nutrients resulting in eutrophication (Hopkinson and Day 1979, 1980a, 1980b; Kemp and Day, 1981; Craig and Day, 1977), a loss in storm buffering capacity, and a loss of important fishery nursery grounds (Turner, 1977 and Chambers, 1980).

Since canals are an important factor affecting land loss, one measure of the impact of canals is potential fisheries loss. An estimated $8-$17 million of fisheries products and services are annually lost in Louisiana due to wetland destruction (Craig et al., 1979). Commercial fish yields are related to the area of coastal wetlands; higher shrimp yields are associated with larger areas of wetlands and only incidentally with water surface area or volume (Turner 1977). Therefore any wetland loss caused by canals is detrimental to fisheries.

Suggested Solutions

1. *Canal Regulation.* Canals are an important agent in wetland altera-tion, affecting not only marsh loss, but salinity intrusion and eutrophication as well. It has been suggested that canaling be permitted only where there is abso-lutely no other alternative and then with a mitigation clause of "no net wetland loss." There are other management techniques that have been suggested to reduce the direct impact of the canals. The least damaging construction tech-nology available could be employed in all cases. Several alternatives are available, including directional drilling that would reduce the number of canals needed for oil and gas exploration, and hydro-air cushion vehicles that would eliminate the need for canals entirely. Mitigation schemes to compen-sate for the unavoidable adverse impacts associated with human activities have also been suggested. Mitigation options that seek to achieve zero habitat loss while maintaining the functional characteristics and processes of wetlands such as natural biological productivity, wildlife habitats, species diversity, water quality, and other unique features have been suggested (Colenen and Cortright, 1979).

2. *Creative Use of Spoil.* Approximately 80% to 90% of the dredging that takes place within the continental U.S. occurs in Louisiana (Lindall and Salo-man, 1977). An enormous amount of spoil is generated every year, and spoil disposal on wetlands is the general rule. It has been reported that for every mile of pipeline installed by flotation canal, 30-36 acres of marsh are altered as a result of spoil deposition (McGinnis et al., 1972).

In order to reverse this trend in wetland alteration, spoil could be viewed as a reusable resource rather than as waste. It is possible to use these sediments productively to create and aid in the management of habitat. A 5-year (1973-1978) Dredged Material Research Program (DMRP) was conducted by the Corps of Engineers at the Waterways Experiment Station in Vicksburg, Mississippi, based on this idea. Several uses of spoil are: a substrate for wildlife habitat such as islands for nesting birds, marsh habitat, beach renewal, restoration of bare ground, construction material, and sanitary landfill. Wetland substrates which are subject to subsidence or erosion can benefit from a deposit of dredged material to replenish what has been lost (National Marine Fisheries 1979; Hunt, 1979). Major field experiments have tested vegetation establishment techniques and principles for spoil. As a result of these studies, the ability to dispose of dredged material in a biologically productive manner was demon-strated, and the engineering characteristics and behavior of dredged and disposed sediments can be predicted and determined.

3. *Barrier Island Stabilization.* Barrier island stabilization has been used to retard land loss of both islands and the wetlands they protect from storm wave activity. Structural and biological approaches have been considered. The structural approach involves construction of groins and riprap, which may stabilize one area at the expense of another. Beach nourishment (pumping sand onto the beach from offshore) is another technique that has been used, especially along the south Atlantic coast. The biological approach generally involves

planting grass to stabilize dunes. This method appears to be successful, at least in the short-term (Mendelssohn, 1982).

4. *Controlled Water Diversions.* Flood control measures such as leveeing along the Mississippi River have interrupted the balance between riverine and marine processes which result in sediment transport, deposition, and introduction of valuable freshwater and nutrients. These processes, which built and stabilized the marsh and swamp areas via overbank flooding, are now virtually eliminated in most of coastal Louisiana.

Schemes for controlled diversions of the Mississippi River have been developed as a means of introducing river water and sediment into wetlands for offsetting wetland loss. "Basically, this approach would reestablish the overbank flow regime of the Deltaic plain, presently disrupted by flood protection levees, and restore more favorable water quality conditions to the highly productive deltaic estuaries" (Gagliano et al., 1981). The feasibility of controlled diversion is indicated by the relatively small input of energy and materials needed to build a major subdelta (Gagliano et al., 1971). The U. S. Army Corps of Engineers (1980) has suggested a number of potential sites for controlled diversions into the Barataria basin. In addition, several diversion schemes for the western Terrebonne basin have been proposed.

The Need for an Integrated Spatial Simulation Modelling Approach

The various suggested solutions to the land loss problem all have far-reaching implications. They depend on which combination of solutions are undertaken and when and where they are undertaken. Outside forces (such as rates of sea level rise) also influence the effectiveness of any proposed solution. In the past, suggested solutions have been evaluated independently of each other and in a "seat of the pants" manner. In order to more objectively evaluate the many interdependent implications of the various management strategies and specific projects that have been suggested to remedy the coastal erosion problem, an integrated spatial simulation modelling approach is needed (Sklar et al., 1985; Costanza et al., 1986). This approach can simulate the past behavior and predict future conditions in coastal areas as a function of various management alternatives, both individually and in any combination. It simulates both the dynamic and spatial behavior of the system, and will keep track of the important variables in the system, such as habitat type, water level and flow, sediment levels and sedimentation, subsidence, salinity, primary production, nutrient levels, and elevation.

Models of this type are large and expensive to implement, but once implemented can be used to quickly and inexpensively evaluate various management alternatives. We envision an ongoing modelling capability for the state's coastal areas, constructed in stages. We have already spent considerable effort designing and implementing this type of model for the western

Terrebonne basin and it has been used to evaluate several current management alternatives for this area.

The Coastal Ecological Landscape Spatial Simulation (CELSS) Model

The model consists of 2,479 interconnected cells, each representing 1 square kilometer. Each cell contains a dynamic simulation model, and each cell is connected to each adjacent cell by the exchange of water and suspended materials. The volume of water crossing from cell to cell is controlled by habitat type, drainage density, waterway orientation, and levee heights. The buildup of land or the development of open water in a cell depends on the balance between net inputs of sediments and organic peats and outputs due to erosion and subsidence. The balance of inputs and outputs is critical, and is important for predicting how marsh succession and productivity are affected by natural and human activities.

Forcing functions (inputs) are specified in the form of time series over the simulation period. Weekly values of Atchafalaya and Mississippi River discharges, Gulf of Mexico salinity, river sediments and nutrients, rainfall, sea level, runoff, temperature, and winds are supplied to the simulation with each iteration. The location and characteristics of the major waterways and levees are also supplied as input to the simulation. Water can exchange with adjacent cells via canals, natural bayous and overland flow or it may be prevented from exchanging with adjacent cells by the presence of levees. The overall water flow connectivity parameter (K2) is adjusted during the model run to reflect the presence and size of waterways or levees at the cell boundaries. If a waterway is present at a cell boundary, a large K2 value is used, increasing with the size of the waterway. If a levee is present, a K2 value of 0. is used until water level exceeds the height of the levee. The model's canal and levee network is updated each year during a simulation run, i.e. dredged canals and levees are added to the model's hydrologic structure at the beginning of the year they were built.

Each cell in the model is potentially connected to each adjacent cell by the exchange of water and suspended materials. Before this exchange takes place however, the ecological and physical dynamics within a cell are calculated. For example, Figure 6-1 shows diagramatically how the water (the large "tank"), suspended sediments (top, small "tank"), and bottom sediments (elevation: bottom, small "tank") components are interconnected within a typical cell. The volume of water crossing from one cell to another carries a specified sediment load. This sediment is deposited, resuspended, lost due to subsidence, and carried to the next cell. The amount of sediment in each "tank" is a function of the habitat type. Not shown in Figure 6-1, but included in the model, is the fact that plants and nutrients within each cell will also influence these exchanges and flows.

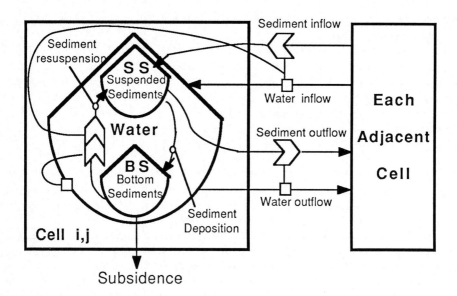

Figure 6-1. Storages (tank symbols) and flows (lines) of water, suspended sediments (SS), and botom sediments (BS) for a typical cell. Fluxes of suspended sediments are a function to water flows and sediment concentrations.

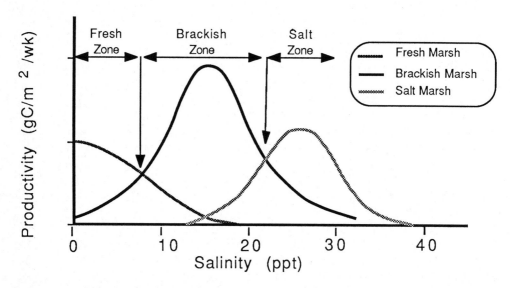

Figure 6-2. Plant primary production as a function of salinity showing points of habitat succession in the model for three of the habitat types.

Habitat succession occurs in the model (after a time lag) when the state variables within a cell become more like another habitat type due to the changing conditions. The biotic components in a cell (primary production) respond to the abiotic changes according to the functions illustrated in Figure 6-2. In this figure, the model's relationship between weekly productivity and salinity for three of the habitat types is shown, indicating that productivity and thus succession change with changing salinity. This same type of function is used to simulate the dependence of primary production upon the cumulative impacts of changing flooding regimes, nutrient levels, turbidity, and elevation.

Model Output

The model can produce a huge amount of output, most useful of which is a contour map for each state variable as well as habitat maps for each week of the simulation. To produce these maps the model must solve over 17,000 simultaneous difference equations and generate over one million simulated data points for each year of simulation. With present day computers models such large numbers of computations are feasible, and as computers continue to improve in speed and convenience, this type of modelling should become more practical.

The results of the model are best comprehended by viewing a video tape of the model's time series mapped output for each state variable and habitat type. Since we can't run the video in this chapter, we present a few "snap shots" of habitat change (Figure 6-3) and discuss some of the findings.

The model predicts the gradual intrusion of salt into the system from the southeastern part of the study area with the concurrent freshening in the northwestern sector. It also illustrates a loss of elevation in the north and an increase in elevation in the south. Both of these trends are indicative of river water and sediments moving further south in recent times plus a lack of connectivity with the more northern fresh marsh areas. Predicted water volume and suspended sediments behaved in a similar way and are generally consistent with what is known about the historical behavior of the system. These physical changes, in turn, have an impact upon the biology of the area. The relationship between plants and elevation of the marsh results in a feedback loop that enhances the rate of land loss as suspended sediments are diverted from an area of marsh.

The model accurately predicted changes in salinity zones and generalized water flow patterns. Overall, the present model does a fairly good job of predicting landscape succession.

To quantitatively assess the validity of the CELSS model, the 1978 habitat map generated by the model is compared to the actual 1978 habitat map. The most straightforward method to compare the maps is to calculate the percent of corresponding cells in the two maps which have the same habitat type. In our current runs, a cell-by-cell comparison has resulted in a fit of 86% correct.

Another method has been proposed (Costanza, in prep) to compare the degree of fit between two maps of categorical data. This method looks at the way the fit

Figure 6- 3. Sample habitat map output from the base case model run (right column) compared with real data (left column).

changes as the resolution of the maps is degraded by using a sampling window of gradually increasing size. The sampling window is moved through the scene one cell at a time until the entire image is covered.

The formula for the fit at a particular sampling window size (F_w) is:

$$F_w = \frac{\sum\limits_{s=1}^{n} \left[1 - \frac{\sum\limits_{i=1}^{p} |a_{1i} - a_{2i}|}{2w^2} \right]_s}{t_w}$$

(6-1)

where:

F_w = the fit for sampling window size w

w = the dimension of one side of the (square) sampling window

a_{ki} = the number of cells of category i in scene k in the sampling window

p = the number of different categories (e.g.,. habitat types) in the sampling windows

s = the sampling window of dimension w by w which slides through the scene one cell at a time

t_w= the total number of sampling windows in the scene for window size w

n= the maximum window size

One can then plot the fit between the scenes (F_w) vs. the size of the sampling window (w) as in Figure 6-4. Figure 6-4 contains plots of goodness of fit, F_w, as a function of window size (w) for various model runs compared with the data. If the plot behaves as it does for the 1978 real vs. simulated maps (i.e., increasing rapidly at small window sizes), the pattern between the two scenes is well matched despite the low fit at window size 1. This would be the case if the patterns between the scenes were similar, but the precise boundaries in the maximum resolution scenes were off. Conversely, if the plot were flat the spatial pattern would not be well matched even though the initial fit might be higher.

A weighted average of the fits at different window sizes seems to be the most appropriate single measure of fit. For this purpose we use the following formula:

$$F_t = \frac{\sum\limits_{w=1}^{n} F_w\, e^{-k(w-1)}}{\sum\limits_{w=1}^{n} e^{-k(w-1)}}$$

(6-2)

This formula gives exponentially less weight to the fit at lower resolution. The value of k determines how much weight is to be given to small vs. large sampling windows. If k is 0 all window sizes are given the same weight. At k = 1 only the first few window sizes will be important. The relative importance of matching the patterns precisely vs. crudely must be answered in the context of the model's objectives and the quality of the data.

Sensitivity analysis is presently being conducted to determine the effects of combinations of over one hundred parameters on the goodness of fit (F_t) of the model. The analysis consists of a factor screen design in which each factor has two levels (high – low). This will provide information on all two-factor interactions in the test as well as imply which three-factor interactions produce the greatest effect in the simulated landscape. Preliminary results indicate that waterflow parameters associated with waterways and habitat type are most sensitive in tuning the model. Further parameter adjustments are expected to achieve an overall spatial goodness-of-fit (F_t) of better than 90%.

The Management Potential of the CELSS Model

The simulation of long-term habitat changes in the coastal marshes of the

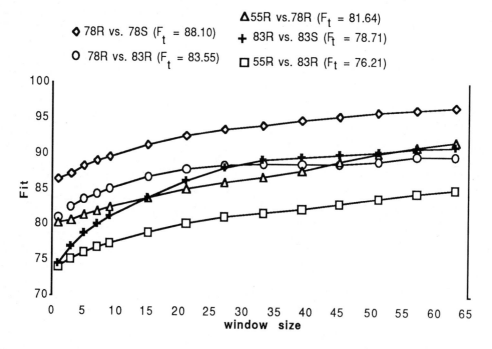

Figure 6-4. Plots of window size (w) vs. fit (F_w) for various runs of the model compared with real data. R refers to real data and S refers to simulated model output. F_t is the total weighted fit over all window sizes.

Atchafalaya River demonstrates that ecological and physical processes can be realistically and relatively accurately modelled. The results of the CELSS Model indicate that the current trend of habitat succession will continue to result in wetland degradation unless something is done. Each oil access canal, levee, and dredge and fill activity that is permitted may seem small and unimportant on a case-by-case basis, appearing only as an insignificant localized impact. However, we have shown that when spatial processes and cumulative impacts are considered, the effects are greatly magnified. In addition, the effects of some management options are contingent upon which other options are simultaneously employed. The long-term implications of canal dredging, for example, may be dependent on the sediment environment, an environment which may be drastically affected by marsh management plans or other options.

We are currently in the process of running the model to the year 2050 for several different scenarios. These include the effects of the proposed Avoca Island levee extension, various canal dredging and backfilling options, controlled water diversions, and semi-impoundments. These options will also be investigated in various combinations. This will give government agency personnel, landowners, oil industry representatives, and the general public better predictions of the complex implications of human activities in the Louisiana coastal zone, and should lead to better management of the resource.

In addition, we can do "hindcasts" with the model to study the impacts of past activities. For example, we have run the model *without* any of the canals dredged between 1956 and 1978 to look at the impacts of past canal dredging. Preliminary results are shown in Figure 6-5. We could also pick out types of

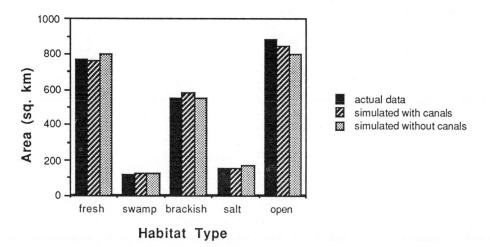

Figure 6-5. Preliminary results of running the model with and without artificial canals and levees. Chart shows the area of each of the habitat types in 1978 for the data and model runs including and excluding canals. Eliminating canals decreases the open water habitat and increases the marsh habitats (except brackish).

canals or even individual canals to eliminate in a simulation and thus estimate historical impacts.

At present, models of this size and complexity are fairly new and expensive. But as we gain experience, and as supercomputers and array processors
become more readily available, models of this type will become practical tools
for understanding and managing natural coastal systems.

Acknowledgments

This research was supported by the U. S. Fish and Wildlife Service. We
thank T. Ozman, S. Leibowitz, P. Nepalli, S. Nellore, and M. Ahmed for their
contributions.

References

Adams, R.D., B.B. Barrett, J.H. Blackmon, B.W. Gane, and W.G. McIntire.
    1976. *Barataria basin: geologic processes and framework.* Louisiana State
    University, Center for Wetland Resources, Baton Rouge, LA. Sea Grant
    Publ. No. LSU-T-76-006.
Baumann, R.H. 1980. *Mechanisms for maintaining marsh elevation in
    subsiding environment.* M.S. thesis, La. State Univ., Baton Rouge, LA.
Baumann, R.H. and R.D Adams. 1981. The creation and restoration of
    wetlands by natural processes in the lower Atchafalaya River system:
    possible conflicts with navigation and flood control management. Pages 1-
    24 in R.H. Stovall, editor. *Proc. 8th Ann. Conf. Wetlands Restoration and
    Creation.* Hillsborough Community College, Tampa, FL.
Baumann, R.H., J.W. Day, Jr., and C.A. Miller. 1984. Mississippi deltaic
    wetland survival: sedimentation vs. coastal submergence. *Science*
    224:1093-1095.
Chambers, D.G. 1980. *An analysis of nekton communities in the upper Bara-
    taria basin, Louisiana.* M.S. thesis, Louisiana State University, Baton
    Rouge, LA.
Cleveland, C.J., C. Neill, and J.W. Day, Jr. 1981. The impacts of artificial
    canals on land loss in the Barataria basin, Louisiana. Pages 425-433 in
    W.J. Mitsch, R.W. Bosserman, and J.M. Klopatek, editors. *Energy and
    Ecological Modelling.* Elsivier, Amsterdam.
Coenen, N.L. and B. Cortright. 1979. Mitigation in the Oregon coastal man-
    agement program. Pages 103-107 in G.A. Swanson, technical coordinator.
    *The Mitigation Symposium: A national workshop on mitigating losses of
    fish and wildlife habitats.* U.S. Dept. Agric., For. Ser. Gen. Tech. Rept.
    RM-65.
Costanza, R. (in preparation). Model goodness of fit: a multiple scale proce-
    dure.
Costanza, R., F. H. Sklar, and J. W. Day, Jr. 1986. Modeling Spatial and Tem-
    poral Succession in the Atchafalaya/Terrebonne Marsh/Estuarine

Complex in South Louisiana. pp. 387-404 in D. A. Wolfe, editor. *Estuarine Variability*, Academic Press, New York.

Craig, N.J., and J.W. Day, Jr. 1977. *Cumulative impact studies in the Louisiana coastal zone: eutrophication and land loss.* Final Report to Louisiana State Planning Office. Louisiana State University, Center for Wetland Resources, Baton Rouge, LA.

Craig, N.J., R.E. Turner, and J.W. Day, Jr. 1979. Land loss in coastal Louisiana (USA). *Environ. Man.* 3:133-144.

Deegan, L.A., H.M. Kennedy, and C.Neill. 1984. Natural Factors and human modifications contributing to marsh loss in Louisiana's Mississippi River deltaic plain. *Environ. Man.* 8:519-528.

DeLaune, R.D., and W.H. Patrick, Jr. 1979. Nitrogen and phosphorous cycling in a Gulf coast salt marsh. Pages 143-151 in V.S. Kennedy, editor. Estuarine Perspectives. Academic Press, NY.

DeLaune, R.D., W.H. Patrick, Jr., and R.J. Buresh. 1978. Sedimentation rates determined by 137Cs dating on a rapidly accreting salt marsh. *Nature* 275:532-533.

Gagliano, S.M. 1973. *Canals, dredging, and land reclamation in the Louisiana coastal zone. Hydrologic and Geologic Studies of the Louisiana Coastal Zone.* Rept. No. 14. Coastal Resources Unit, Center for Wetland Resources, Louisiana State University, Baton Rouge.

Gagliano, S.M., K.J. Meyer-Arendt, and K.M. Wicker. 1981. Land loss in the Mississipi River Deltaic Plain. *Trans. Gulf Coast Assoc. Geol. Soc.* 31:295-300.

Gagliano, S.M., P.P. Light, and R.E. Becker. 1971. *Controlled diversion in the Mississippi River delta system: an approach to environmental management.* Hydrologic and Geologic Studies of the Louisiana Coastal Zone, Rept. No. 8. Coastal Resources Unit, Center for Wetland Resources, Louisiana State University, Baton Rouge.

Hopkinson, C.S., and J.W. Day, Jr. 1977. A model of the Barataria Bay salt marsh ecosystem. Pages 235-266 in C.A.S. Hall and J.W. Day, Jr., editors. *Ecosystem Modelling in Theory and Practice: An Introduction with Case Histories.* John Wiley, NY.

Hopkinson, C.S., and J.W. Day, Jr. 1979. Aquatic productivity and water quality at the upland-estuary interface in Barataria basin, Louisiana. Pages 291-314 in R. Livingston, editor. *Ecological Processes in Coastal and Marine Systems.* Plenum Press, NY.

Hopkinson, C.S., and J.W. Day, Jr. 1980a. Modelling the relationship between development and storm water and nutrient runoff. *Environ. Man.* 4:315-324.

Hopkinson, C.S., and J.W. Day, Jr. 1980b. Modelling hydrology and eutrophication in a Louisiana swamp forest ecosystem. *Environ. Man.* 4:325-335.

Hunt, L.J. 1979. Use of dredged materials disposal in mitigation. Pages 502-507 in G.A. Swanson, technical coordinator. *The Mitigation Symposium: A National Workshop on Mitigating Losses of Fish and Wildlife Habitat.*

General Technical Report RM-65.

Kemp, G.P., and J.W. Day, Jr. 1981. *Floodwater nutrient processing in a Louisiana swamp forest receiving agricultural runoff.* Louisiana Water Resources Research Institute, Louisiana State University, Baton Rouge. Report No. A-043-LA. 60 pages.

Lindall, W.N., Jr., and C.H. Saloman. 1977. Alteration and destruction of estuaries affecting fishery resources of the Gulf of Mexico. *Marine Fisheries Review* 39:1-7.

McGinnis, J.T., R.A. Ewing, C.A. Willingham, S.E. Rogers, D.H. Douglass, and D.L. Morrison. 1972. *Final report on environmental aspects of gas pipelines in marshes.* Battelle, Columbus Lab, 505 King Avenue, Columbus, OH.

Mendelssohn, I.A. 1982. Sand dune vegetation and stabilization in Louisiana. Pages 187-207 in D.F. Boesch, editor. *Proc. Conference on Land Loss and Wetland Modifications in Louisiana: Causes, Consequences, and Options.* U.S. Fish and Wildlife Service Report FWS/OBS-82/59. Washington, DC.

Morgan, J.P., and P.B. Larimore. 1957. Changes in the Louisiana shoreline. Trans. Gulf Coast Assoc. Geol. Soc. 7:303-310.

National Marine Fisheries Service. 1979. *Guidelines and criteria for proposed setland alterations in the southeast region.*

Scaife, W.W., R.E. Turner, and R. Costanza. 1983. Coastal Louisiana recent land loss and canal impacts. *Environ. Man.* 7:433-442.

Sklar, F. H., R. Costanza, and J. W. Day, Jr. 1985. Dynamic spatial simulation modeling of coastal wetland habitat succession. *Ecol. Modeling.* 29:261-281.

Turner, R.E. 1977. Intertidal vegetation and commercial yields of penaeid shrimp. *Trans. Am. Fish. Soc.* 106:411-416.

U.S. Army Corps of Engineers. 1980. *New Orlean–Baton Rouge metropolitan water resources study.* U.S. Army Engineer District, New Orleans, LA.

Van Sickle, V.R., B.B. Barrett, L.J. Gulick, and T.B. Ford. 1976. *Barataria basin: salinity changes and oyster distribution.* Center for Wetland Resources, Louisiana State University, Baton Rouge, LA. Sea Grant Publ. LSU-T-76-002 and Louisiana Department of Wildlife and Fisheries Tech. Bull. 20:1-22.

7/ PRODUCTIVITY-HYDROLOGY-NUTRIENT MODELS OF FORESTED WETLANDS

William J. Mitsch

Modelling of forested wetlands depends on an understanding of the effects of hydrology and nutrient conditions on the primary productivity of the wetland ecosystem. Several studies, done over a decade and a half in the eastern half of the United States, suggest that flow-through forested wetlands are most productive and sluggish forested wetlands the least. Some investigators have attempted to translate these general findings into mathematical statements or other quantitative relationships. These quantifications include statistical relationships of primary productivity as a function of hydrology and/or nutrient inflow, and parabolic curves depicting productivity as a function of a hydrologic variable (i.e. water depth, depth to water table, or flow-through conditions). A more detailed summary is given on a several-year project in western Kentucky, U.S., where determination of effects of wetland hydrodynamics and nutrient conditions on forested wetland productivity was a primary research goal. A preliminary forested wetland model, which simulates the influence of hydrology and nutrient conditions on wetland productivity, illustrates highest productivity with pulsing hydrology and least with sluggish, low nutrient conditions.

Introduction

The application of ecological modelling to the study of forested wetlands is only in its early phases of development. One reason for the previous lack of model development for forested wetlands is that, until recently, there was little recognition of any management values for these systems except for timber harvest. Forested wetlands, which predominate in the southeastern United States but are found along rivers throughout the eastern half of North America, are now recognized for a number of other possible values, including fish and wildlife protection, water quality improvement, and flood prevention. Figure 7-1 illustrates these management alternatives as well as several others such as hydrologic modification and wetland removal. All of these possible functions of forested wetlands suggest that these systems are good candidates for ecological modelling approaches to assist in assessing management alternatives.

A second reason for the limited amount of modelling of forested wetlands is the general poor understanding of how these systems function. This is particularly true for our understanding of the functional relationships between a forested wetland's hydrology, which is one of its most important ecological signatures, and the wetland ecosystem response. For example, unlike deepwater aquatic systems which are dependent primarily on the availability of nutrients for the levels of primary productivity, forested wetlands are dependent on hydrologic conditions as well as nutrients for their productivity. Nutrient conditions are also very important in determining forested wetland productivity, but only insofar as the hydrologic conditions control the chemical ones.

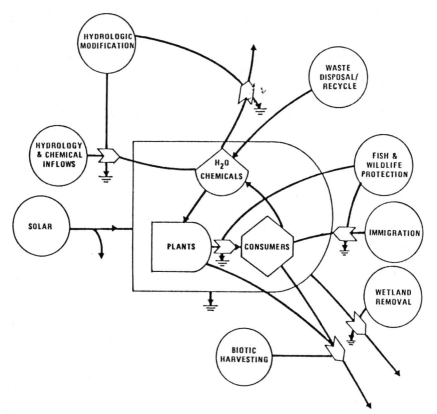

Figure 7-1. Conceptual model of forested wetland management alternatives.

This paper will first review and discuss studies on the general relationships among hydrology, nutrients, and primary productivity of forested wetlands as precursors of more detailed simulation models. The author will then present a preliminary simulation model, based in part on a research project of several years in western Kentucky, that demonstrates theories about relationships among hydrology, nutrients, and wetland processes.

Forested Wetland Productivity Models

Several investigators have undertaken studies to compare forested wetland productivity under different hydrologic regimes. While only a few of these studies have quantified the relationship between hydrologic variables and wetland productivity, they represent the first step toward developing sub-models of forested wetland productivity, particularly as it relates to hydrology. Table 7-1 summarizes some of the studies of forested wetland productivity which will be discussed here.

Establishing the Hydrologic Relationship

Carter et al. (1973) were among the first investigators to demonstrate the effect of altered hydrologic conditions on productivity of forested wetlands. Their findings showed that overall productivity forested cypress strand wetlands in Florida decreased significantly when the strands were ditched and drained. A subsequent study by Conner and Day (1976) in Louisiana demonstrated the difference in productivity between a continuously flooded *Taxodium-Nyssa* swamp and a seasonally-flooded bottomland hardwood wetland. Those results, summarized in Table 7-2, demonstrated that a pulsing hydroperiod rather than a continuously flooded hydroperiod lead to a greater net primary productivity of forested wetlands. In a follow-up to that study, Conner, Gosselink, and Parrondo (1981) found high productivity in forested *Taxodium-Nyssa-Fraxinus* wetlands that were impounded part of the year, drained in the growing season, and often flooded with freshwater to ensure adequate oxygen. Again, lowest productivities were found in stagnant, continuously flooded forested wetlands.

Mitsch and Ewel (1979) investigated tree growth data from a number of sites in central Florida and found that cypress-hardwood associations, with their species composition verifying their pulsing hydrologic conditions, had the most productive cypress trees, while stagnant monospecific cypress stands had the lowest productivity. They found that cypress trees in cypress-pine associations, an indication of dry conditions, also had slow growing cypress trees. Brown (1981) found a similar pattern of productivity for flow-through, stillwater, and stagnant cypress wetlands in Florida and concluded that nutrient inflow, which is coupled to hydrologic inflow, is the important variable which determines productivity. These studies and others led Brinson et al. (1981) to summarize the net biomass productivity of forested wetlands in order of greatest to least productivity as:

flowing water swamp > sluggish flow swamp > stillwater (stagnant) swamp

Table 7-1. Studies of comparative productivity of forested wetlands in eastern United States.

Location	Study	Reference
Southern Florida	Drained vs. Undrained Swamps	Carter et al. 1973
Louisiana	Bottomland Hardwood vs. *Taxodium - Nyssa* Swamp	Conner and Day, 1976
General	Riparian Wetland Productivity	Odum, 1979
Central Florida	Several *Taxodium* Associations	Mitsch and Ewel, 1979
Southern Illinois	Tree Ring Growth vs. Flooding for *Taxodium* Swamp	Mitsch et al., 1979
Florida	Carbon Metabolism Studies for *Taxodium* Swamps	Brown, 1981
Louisiana	Impounded and Natural Swamps	Conner et al., 1981
General	Comparison of Literature	Gosselink et al., 1981; Conner and Day, 1982
Northeastern Illinois	Tree Ring Growth vs. Flooding Bottomland Forest Trees	Mitsch and Rust, 1984
Western Kentucky	Productivity of Forested Wetlands with Different Hydroperiods	Taylor, 1985; Benson, 1986

Table 7-2. Comparison of net primary productivity of forested wetlands in Louisiana in g/m2-yr (from Conner and Day, 1976 and Conner et al., 1981).

Measurement	Pulsing Managed	Pulsing Natural	Permanent Flooded	Permanent Stagnant
Stem Increase	1230	800	500	560
Leaf Litterfall	550	574	620	330
Understory Prod.	--	200	20	--
Total NPP	**1780**	**1574**	**1140**	**890**

Quantifying the Relationship

The relationship between forested wetland productivity and hydrology has only begun its initial stages of quantification from these and similar studies, and that quantification has been primarily through the development of graphs, some with actual data and some as only theoretical plots. Several of these relationships are shown in Figure 7-2. Gene Odum (1979) was one of the first to suggest a parabolic curve to describe wetland productivity versus flooding regime. This diagram, shown in Figure 7-2a, emphasizes that the seasonal flooding can be viewed as a a subsidy to the system while both stagnant conditions and abrasive flooding are more stress than subsidy to the forested wetland. This curve, seen in several other investigators' works, is a fundamental one in ecology, dating back to Shelford's limitation curves. Mitsch and Ewel (1979), in summarizing their measurements of *Taxodium* (cypress) growth in a variety of plant associations (and therefore in a variety of hydrologic conditions), concurrently produced a similar parabolic curve to describe their theory of hydrologic effects on cypress in particular and forested wetlands in general (Figure 7-2b). Highest productivities were suggested to be in the conditions where there were greatest differences between dry and wet seasons (pulsing), while low productivities result when the system is either too wet (stagnant) or too dry (invasion of upland species).

Mitsch et al. (1979), in describing the growth of cypress trees for dry and wet seasons in a forested wetland in southern Illinois, established the following mathematical regression for the growth of *Taxodium* as a function of annual flooding (Figure 7-2d):

$$G_{ba} = 24 + 0.43\,Q_{ave} \qquad\qquad (7\text{-}1)$$

where,

G_{ba} = basal area growth of individual trees, cm^2/yr

Q_{ave} = average river discharge, cm/yr

It is clear that this relationship is quite specific to an exact baldcypress swamp in southern Illinois and to the hydrologic conditions of the adjacent Cache River. That paper also showed that there were similar statistically significant relationships for growth of trees when regressed against number of floods per year or the total of the flood peaks. Statistical relationships such as these have little application for other sites, yet this equation represents one of the first quantitative empirical descriptions of the hydrology—productivity relationship.

Brown (1981), by comparing several studies from the southeastern United States, established a relationship between forested wetland productivity and phosphorus inflow as a measure of hydrologic and nutrient flux through the wetland:

$$NBP = 1012 + 143\,(\ln P_{in}) \qquad\qquad (7\text{-}2)$$

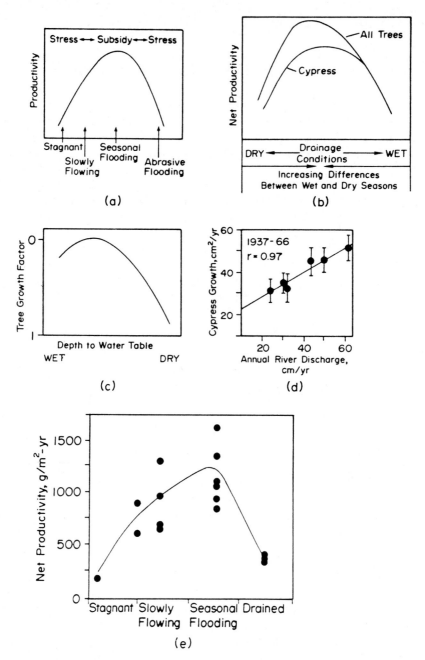

Figure 7-2. Relationships between hydrology and forested wetland productivity as presented by different investigators: a) general relationship as suggested by Odum (1979); b) productivity vs. water regime for forested wetlands in Florida (from Mitsch and Ewel, 1979); c) tree growth as a function of depth to water for bottomland hardwood forest model (from Phipps, 1979); d) cypress tree growth as a function of flooding (Mitsch et al., 1979); e) summary graph of net primary productivity measurements as a function of hydrologic conditions (Conner and Day, 1982).

where,

NBP = net biomass production of forested wetland, g/m^2-yr

P_{in} = phosphorus inflow, g P/m^2-yr

The interesting result of this study was the illustration, through the comparison of data from several wetland sites, that the productivity not only depended on the hydrologic conditions of the wetland, but also on the nutrient conditions. Brown chose to use phosphorus inflow as an independent variable in predicting primary productivity because this term implicitly includes both hydrologic inflow and the nutrient conditions of the surrounding (or upstream) watershed.

Phipps (1979) and Phipps and Applegate (1983) applied an upland forest model originally developed by Bodkin et al.(1972) to southern United States forested wetlands. They developed a submodel that described the relative growth of trees as a function of depth to the water table, i.e. as conditions go from wet to dry. This relationship, too, was given as a parabolic function (Figure 7-2c) and is expressed in equation form as:

$$G_i = G_{MAXi} (1 - 0.055 (T - W_i)^2) \tag{7-3}$$

where,

G_i = productivity or growth of tree species i

G_{MAXi} = maximum productivity or growth of species i

T = depth of water table

W_i = optimum water table depth for species i

Pearlstein, McKellar and Kitchens (1985) investigated the impact of altered hydrologic regime on bottomland forests of South Carolina with a successional model called FORFLO. In that model, hydrologic parameters were used to control seed germination, tree growth, and tree mortality. That model used a variation of the relationship developed by Phipps (1979) to describe the growth of the wetland trees as a function of water levels:

$$G_i = G_{maxi} (1.0 - 1.3 (T - W_i)^2) \tag{7-4}$$

where,

G_i = growth of individual tree of species i

G_{maxi} = optimum growth of species i

T = water level (or water table depth)

W_i = optimum water table depth for species i

Several investigators have chosen to summarize the data with general parabolic functions, showing both empirical and theoretical relationships. One such summary was presented by Conner and Day (1982), who showed the results of several of these forested wetland studies in a graphic format (Figure 7-2e), with net primary productivity quantified on the abscissa (ranges from ca. 200 to 1750 grams/m²-yr) but with the hydrologic conditions along the ordinate only qualitatively presented as stagnant (continually flooded), slowly flowing, seasonal, and drained. What remains in order to develop proper ecological models of forested wetlands is to quantify these hydrologic conditions, perhaps in terms of water depth, flooding duration and frequency, and water renewal rate (inflows/average water storage).

It is clear from these studies that quantitative approaches to understanding the relationships among productivity, nutrients, and hydrology are only now beginning to be developed. Some progress has been made to describe the relationships for individual trees (see Phipps and Pearlstein models in Equations 7-3 and 7-4). Describing whole ecosystem response to hydrology and nutrients has been much less rigorous in mathematical descriptions.

A Model of Forested Wetlands

Background of Study

Forested wetlands in western Kentucky have been the subject of systems investigations by the author and his students and colleagues (Mitsch et al., 1981; Mitsch et al.,1983a,b; Cardamone et al., 1984; Taylor, 1985; Benson, 1986). Specific sites for detailed study of forested wetland ecosystem dynamics are shown in Figure 7-3. One objective of these studies was to use models to guide the research efforts and to develop data that could be used in developing simulation models. Wetland productivity relationships, as studied by Taylor (1985), and nutrient dynamics, as studied by Benson (1986), demonstrated that there is considerable dependence of wetland productivity on hydrologic conditions and nutrient inflow. Data from those studies, summarized in Table 7-3, demonstrate that the forested wetlands which were flooded for less that 30 percent of the year, and which had average depths below the ground surface, were the most productive (1280 and 1334 g/m²-yr). These wetlands were flooded, primarily in the spring, one by a small river, Cypress Creek, the other by the large Ohio River. Two wetlands on the Ohio River floodplain which were slowly flowing and were flooded 90 to 100 percent of the time were next lowest in productivities (634 and 524 g/m²-yr). The lowest productivity was measured at a stagnant forested wetland dominated by *Taxodium*, which was on the floodplain of Cypress Creek, but impounded and isolated from the creek. Here the total productivity was only 205 g/m²-yr. Phosphorus inflows were estimated for the three deeper wetlands, and these estimates suggested a definite positive relationship between wetland canopy productivity and phosphorus loading as originally proposed by Brown (1981). The results also suggest a difference in productivity of forested wetlands based on the size of their resident watershed. The forested wetlands along the Ohio River had generally higher productivity than did the wetlands along the much smaller Cypress Creek. Wetlands along the Ohio River were subject to large annual floods, usually in the spring, while flooding

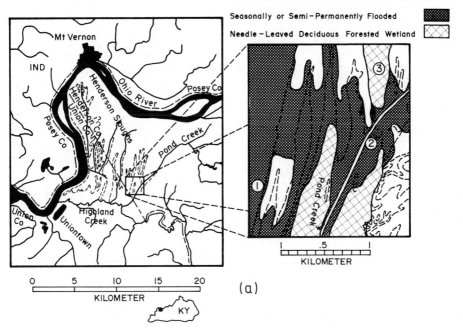

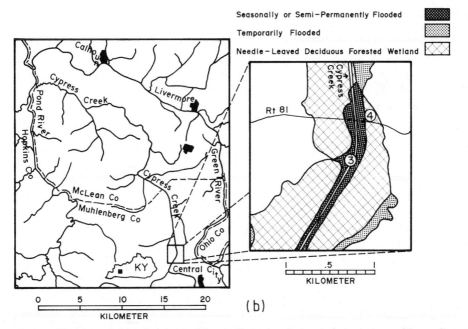

Figure 7-3. Locations of study areas of forested wetland research in western Kentucky: a) Henderson Sloughs, Henderson County, Kentucky, and b) Cypress Creek, Muhlenberg County, Kentucky.

Table 7-3. Hydrology, nutrients, and productivity of five forested wetlands in western Kentucky (from Taylor, 1985 and Benson, 1986).

Parameter	Site Number[a]				
	H1	C3	H2	C4	H3
Flood Hydroperiod	Seasonal	Seasonal	Semi Permanent	Semi Permanent	Permanent
Flow Condition When Flooded	Flowing	Flowing	Slow Flow	Stagnant	Slow Flow
Percent of Growing Flooded	16.7	<30	90	95	100
Percent of Year Flooded	18.8	<30	95	95	100
Ave Depth, m	-0.4	-0.3	0.6	0.5	1.1
Max Water Depth, m	3.4	>1	3.7	<2	3.8
Stem Growth, g/m^2-yr	914	812	498	142	271
Litterfall[b], g/m^2-yr	420	468	136	63	253
Net Primary Productivity, g/m^2-yr	1,334	1,280	634	205	524
Net Runoff into Wetland[c], cm/yr	---	---	160	40	95
Renewal Rate, yr^{-1}	---	---	4.3	2.8	1.8
Net Phosphorus Inflow, gP/m^2-yr	---	---	0.18	0.02	0.08

[a] Large River sites at Henderson Sloughs (H) near Ohio River; Small River sites near Cypress Creek (C).

[b] Litterfall includes leaves and fruit

[c] Net runoff and phosphorus inflow estimated with hydrologic budget that assumed no groundwater interchange.

occurred on Cypress Creek with greater frequency but for shorter duration. Although the data cannot determine this conclusively, there is evidence that wetlands along large river systems are more productive than are equivalent wetlands along smaller streams. This is probably due to a combination of hydrology and higher nutrient conditions in the larger watershed.

A Preliminary Model

A model was developed to illustrate the effects of hydrology and nutrients on forested wetland productivity as shown in the previously described field studies and as described somewhat qualitatively in several previous studies at other sites. The model, shown in Figure 7-4, has simple first order differential equations to simulate water, nutrients (here chosen as phosphorus, most often believed to be limiting in freshwater forested wetlands), and ecosystem biomass. Parameters and differential equations used in the model are shown in Table 7-4. The model was simulated with the microcomputer simulation program STELLA, using 4th order Runge-Kutta integration, a time step of 0.1 year, and a simulation time of 100 years. The model was run for low nutrient and high nutrient conditions and for hydrologic conditions of stagnant (total inflows = 200 cm/yr), flowing (total inflows = 2,000 cm/yr), and pulsing (inflow of 1,000 cm/yr plus a pulse of 1,000 cm/yr to simulate a once-a-year flood).

Examples of simulation results are shown in Figure 7-5 for biomass in high nutrient watersheds and under the three hydrologic conditions. Clearly, the highest productivities resulted with the pulsing system, the next lowest in the flowing system, and the lowest in the stagnant conditions. The renewal rate was maintained the same for all of these simulations (t^{-1} = 1.0/yr). The flow system had higher ultimate productivity because more nutrients were being brought into the system and the wetland was retaining more. The pulsing system had the highest productivity because of high nutrients and because the pulsing hydrology allowed the water level to decrease between floods to levels closer to the optimum requirements of the vegetation. The optimum water level for the vegetation was assumed to be 1.0 m below the soil surface, a reasonable assumption based on data on optimum depth of forested wetland trees provided by Pearstein et al. (1985).

Conditions in each wetland after simulated time of 100 years are given in Figure 7-6. Biomass, productivity, and nutrients all followed the pattern of highest values in the pulsing system, next lowest in the flowing system, and lowest in the stagnant conditions. One exception occurred in the low nutrient conditions, where nutrient conditions were so low in the surrounding watershed (P_{conc} = 0.1 mg/l) that nutrients did not increase from stagnant to flowing conditions, and hence productivity was the same under both conditions. A seventh simulation, where two flood pulses were introduced per year, showed the highest biomass accumulation after 100 years. Net primary productivity, measured at the end of this simulation, was slightly lower that the

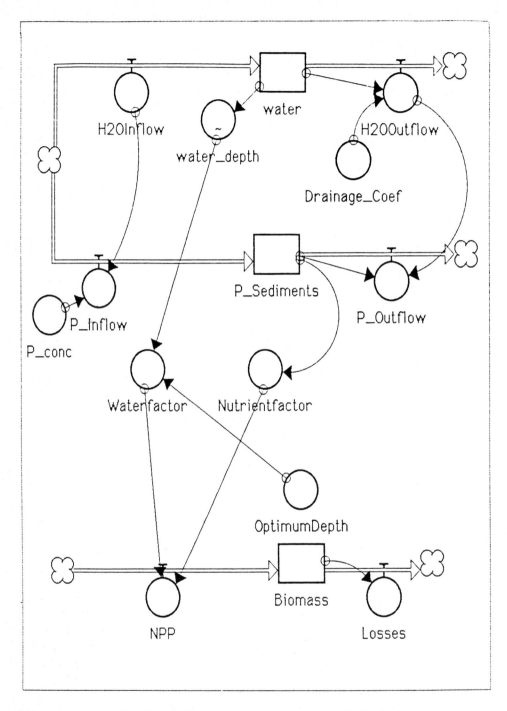

Figure 7-4. Simulation model of forested wetlands, developed with STELLA software for Macintosh microcomputer. State variables are in squares, auxiliary variables in circles, flow of materials in bold lines, and flow of information in single lines.

Table 7-4. Equations, rates, and parameters used in forested wetland model shown in Figure 7-4.

Differential Equations

Water (m)

$$dQ_1/dt = \text{Inflow - Outflow}$$

$$
\begin{aligned}
\text{Inflow} &= 2 \text{ m/yr} \quad \text{(stagnant)} \\
&= 20 \text{ m/yr} \quad \text{(flowing)} \\
&= 10 \text{ m/y} + \text{Pulse (10 m)} \quad \text{(pulsing)} \\
\text{Outflow} &= k_1 * Q_1
\end{aligned}
$$

Phosphorus in Root Zone (g/m²)

$$dQ_2/dt = P_{conc} \text{ Inflow} - k_2 \; Q_2 \text{ Outflow}$$

$$
\begin{aligned}
P_{conc} &= \text{average concentration of phosphorus} \\
&= 0.1 \text{ mg/l for low nutrient systems} \\
&= 1.0 \text{ mg/l for hign nutrient systems}
\end{aligned}
$$

Biomass (kg/m2)

$$dQ_3/dt = \text{NPP - Losses}$$

$$
\begin{aligned}
\text{NPP} &= \text{NPP}_{max} \text{ (Water factor) (Nutrient factor)} \\
\text{Losses} &= k_3 \; Q_3 \\
\text{Water factor} &= 1.0 - 0.05(D_{opt}-D)^2 \\
\text{Nutrient factor} &= Q_2/(k_4+Q_2)
\end{aligned}
$$

Parameters

$k_1 = 1.0 \text{ yr}^{-1}$

$k_2 = .0002$

$k_3 = .05 \text{ yr}^{-1}$

$k_4 = 300 \text{ g/m}^3$

$\text{NPP}_{max} = 2.0 \text{ kg/m}^2\text{-yr}$

$D_{opt} = -1.0 \text{ m}$

D = equivalent "depth" of water in soil (-) or on surface (+)

$\quad = (Q_1/\mu) - 3.0 \quad 0.0 < Q1 < 3\mu$

$\quad = Q_1-3\mu \quad 3\mu < Q1 < 1 +3\mu$

$\quad = 1.0 \quad Q1 > 1 + 3\mu$

μ = soil porosity

$\quad = 0.5$

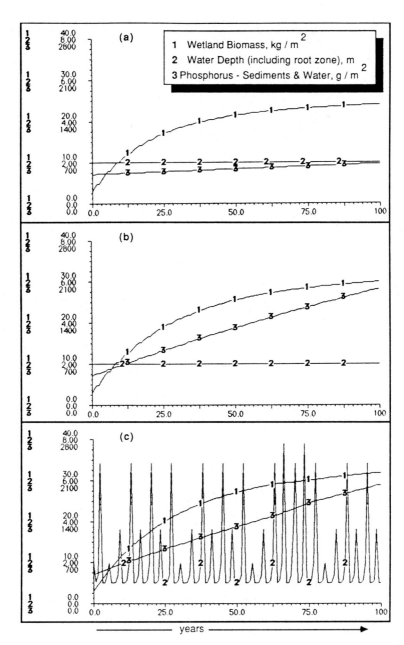

Figure 7-5. Examples of simulation results for simulation model shown in Figure 7-4. Results are for high nutrient watersheds (average phosphorus concentration in water is 1.0 mg-P/l) and show wetland biomass (kg/m2), water storage in root zone and surface (meters), and phosphorus storage in sediments and water (g/m2). Results are for a) stagnant conditions (flowthrough = 2 m/yr); b) flowing conditions (flowthrough = 20 m/yr); and c) pulsing conditions (flowthrough = 10 m/yr plus annual pulse of 10 m).

simulation with one flood per year, only because it is an instantaneous reading by the computer at the end of year 100 and does not represent a slightly higher productivity when the total productivity is integrated over 100 years.

The model simulation results are compared in Figure 7-7, where net biomass production (this is the biomass produced over 100 years, divided by 100 years and does not represent the true net biomass productivity at year 100) is plotted versus phosphorus loading. Otherwise the results are similar to the relationship developed by Brown (1981) in Equation 7-2. This diagram illustrates the importance that nutrient concentrations, as well as overall loading, have on system productivity. Pulsing floods have a more impact on productivity in high nutrient watersheds.

Conclusions

These simulation results should be viewed as only those from a preliminary model, and the absolute values of the results may be difficult to observe in nature. Future models should include more detail of hydrologic conditions, including a more accurate picture of the hydroperiod of an array of wetland types than simply as "stagnant," "flowing," and "pulsing." Nutrient conditions were simplified into "low" and "high" descriptions. Further analysis of forested wetlands and the chemistry of their watersheds may prove that it may be more important to describe the wetlands as "clay" or "silt" dominated watersheds, and to describe the nutrients carried into the wetland as available or non-available (as particulate and/or organic P). The model says nothing about the role of detritus on nutrient retention and neither uptake nor recycling of nutrients by the plants is considered. This may be fairly reasonable however, as the trees have a considerable supply of nutrients, and uptake or recycling should not have any significant impact on the storage of available nutrients. The model does not include other biotic components of the ecosystem, although the biomass storage conceivably includes all of the biotic components. Some disaggregation may be appropriate here. Finally, the model shows no export of organics produced by the "open" wetland, where an obvious drain of nutrients (although of the unavailable form) would occur at least during flooding season. The model does sacrifice these specifics, but it does support, with general numbers substituted for site-specific realism, the overall concepts of hydrology and productivity elucidated by several prior empirical studies. The model demonstrates how modelling can be used with the theory already developed in a decade of studies to "experiment" with wetlands in a way that would be impossible to observe in field studies.

Acknowledgement

This paper was supported, in part, by the University of Louisville, The Ohio State University and a Fulbright Grant to the author to Denmark in 1986-1987. Research in Kentucky was supported by Kentucky Water Resources Institute and U.S. Department of Interior.

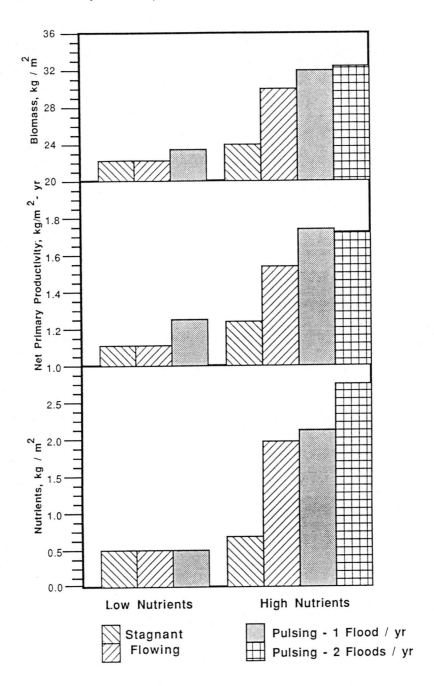

Figure 7-6. Simulation results for biomass, net primary productivity, and nutrients of forested wetland model for low and high nutrient conditions and various hydrologic conditions. Results shown are after simulated 100 year development of wetland which started with same initial conditions.

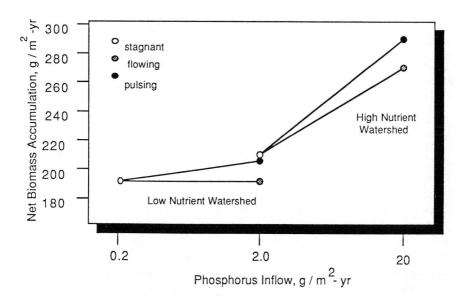

Figure 7-7. Simulation results showing net biomass increase after 100 years versus phosphorus inflow. Low and high nutrient watershed refers to general nutrient conditions in water. Stagnant, flowing, and pulsing conditions are described in Table 7-4.

References

Benson, K.B. 1986. *Hydrologic patterns and phosphorus dynamics in forested wetlands of western Kentucky.* Ph.D. Dissertation, University of Louisville, Louisville, Kentucky, 149 pages.

Bodkin, D.B., J.F. Janak, J.R. Wallis. 1972. Some ecological consequences of a computer model of forest growth. *J. Ecol.* 60:849-872.

Brinson, M.M., A.E. Lugo, and S. Brown. 1981. Primary productivity, decomposition, and consumer activity in freshwater wetlands. *Annu. Rev. Ecol. Systematics* 12:123-161.

Brown, S.L. 1981. A comparison of the structure, primary productivity, and transpiration of cypress ecosystems in Florida. *Ecol Monogr.* 51:403-427.

Cardamone, M.A., J.R. Taylor, and W.J. Mitsch. 1984. *Wetlands and coal surface mining: a management handbook.* Kentucky Water Resources Research Institute Report 154, Lexington, Kentucky, 99 pages.

Carter, M.R., L.A. Burns, T.R. Cavinder, K.R. Dugger, P.L. Fore. D.B. Hicks, H.L. Revells and T.W. Schmidt. 1973. *Ecosystem analysis of the Big Cypress Swamp and estuaries.* US. EPA 904/9-74-002, Region IV, Atlanta, Georgia.

Conner, W.H. and J.W Day, Jr. 1976. Productivity and compostion of a bald cypress—water tupelo site and a bottomland hardwood site in a Louisiana swamp. *Amer. J. Bot.* 63:1354-1364.

Conner, W.H. and J. W. Day, Jr. 1982. The ecology of forested wetlands in the southeastern United States. Pages 69-87 in B. Gopal, R.E. Turner, R. G.

Wetzel, and D.F. Whigham, editors. *Wetlands: Ecology and Management*. National Institute of Ecology and International Scientific Publications, Jaipur, India.

Conner, W.H., J.G. Gosselink, and R.T. Parrondo. 1981. Comparison of the vegetation of three Louisiana swamp sites with different flooding regimes. *Amer. J. Bot.* 68:320-331.

Gosselink, J.G., S.E. Bayley, W.H. Conner, R.E. Taylor. 1981. Ecological factors in the determination of riparian wetland boundaries. Pages 197-219 in J.R. Clark and J. Benforado, editors. *Wetlands of Bottomland Hardwood Forests*. Elsevier, Amsterdam.

Mitsch,W.J., C.L. Dorge, and J.R. Wiemhoff. 1979. Ecosystem dynamics and a phosphorus budget of an alluvial cypress swamp in southern Illinois. *Ecol.* 60:1116-1124.

Mitsch, W.J. and K.C. Ewel.1979. Comparative biomass and growth of cypress in Florida wetlands. *Amer. Midl. Nat.* 101:417-426.

Mitsch, W.J., R. W. Bosserman, P.L. Hill, J.R. Taylor and F. Smith. 1981. Models of wetlands amid surface mining regions of western Kentucky. Pages 103-113 in W. J. Mitsch, R.W. Bosserman, and J.M. Klopatek, editors. *Energy and Ecological Modelling*, Elsevier, Amsterdam.

Mitsch, W.J., J.R. Taylor, K.B. Benson, and P.L. Hill, Jr. 1983a. *Atlas of wetlands in the principle coal surface mine region of western Kentucky*, U.S. Fish & Wildlife Service Report FWS/OBS 82/72, Washington D.C., 135 pages.

Mitsch, W.J., J. R. Taylor, K.B. Benson, and P.L. Hill, Jr. 1983b. Wetlands and coal surface mining in western Kentucky—a regional impact assessment. *Wetlands* 3:161-179.

Mitsch, W.J. and W.G. Rust. 1984. Tree growth responses to flooding in a bottomland forest in northeastern Illinois. *Forest Sci.* 30:499-510.

Odum, E.P. 1979. Ecological importance of the riparian zone. Pages 2-4 in R.R. Johnson and J.F. McCormick, tech. coords., *Strategies for protection and management of floodplain wetlands and other riparian ecosystems*. U.S. Forest Service General Technical Report WO-12, Washington, D.C.

Pearlstein, L., H. McKellar, and W. Kitchens. 1985. Modelling the impact of a river diversion on bottomland forest communities in the Santee River floodplain, South Carolina. *Ecol. Modelling.* 29: 283-302.

Phipps, R.L. 1979. Simulation of wetlands forest vegetation dynamics. *Ecol Modelling.* 7:257-288.

Phipps, R.L. and L.H. Applegate. 1983. Simulation of management alternatives in wetland forests. Pages 311-339 in S.E. Jorgensen and W.J. Mitsch, editors. *Application of Ecological Modelling in Environmental Management, Part B.* Elsevier, Amsterdam.

Taylor, J.R. 1985. *Community structure and primary productivity in forested wetlands in Western Kentucky*. Ph.D. dissertation, University of Louisville, Louisville, Kentucky, 139 pages.

8/ MODELLING NUTRIENT RETENTION BY A REEDSWAMP AND WET MEADOW IN DENMARK

Sven E. Jørgensen
Carl C. Hoffmann
William J. Mitsch

Simulation modelling is suggested as a management tool for investigating the usefulness of wet meadows and other wetlands in Denmark for the protection of surface and ground waters from nutrient inflows from agricultural lands. Field studies are currently measuring many physical, chemical, and biological aspects of a reedswamp adjacent to a lake and a wet meadow which is between a stream and agricultural upland. A modelling effort is being incorporated into these research projects to assist in the design of experimental measurements and to determine the efficiency of using wetland systems as buffer zones between agriculture and aquatic systems.

Introduction

Non-point sources of nutrients are becoming more and more recognized as significant contributors to the eutrophication of lakes, fjords and bays. The use of wetlands as nutrient traps is one of the few available methods to reduce non-point nutrient discharges. Therefore, there has been an increasing interest in modelling the cycling of nutrients in wetlands with the following objectives: 1) to understand the function of these ecosystems in relation to their nutrients cycles; 2) to be able to quantify their nutrient removal capacity under various conditions; and 3) to examine whether it would be feasible to operate wetlands for optimum nutrient removal.

Two wetlands are being investigated in Denmark for their nutrient removal capabilities—a reedswamp and a wet meadow. The features of these two wetlands are given below, together with the results of the investigations and a general model. Only limited results of the modelling effort have been achieved at this stage, and they are given after the presentation of the model. The last section of the chapter contains conclusions that can be made at this stage and recommendations for modelling nutrient dynamics in wetlands. More comprehensive results and conclusions will most probably be achieved when the projects are completed in the coming years.

Description of the Two Wetland Studies

The Reedswamp

It became clear from the development of a eutrophication model for the shallow Lake Glumsø in Denmark (see Chapter 10) that non-point sources of nutrients are very significant, mainly due to intensive agriculture in the catchment area. That raises the question: Are we able to control these sources? At the inflow of the main tributary to Lake Glumsø is a reedswamp. Would it be possible to control the non-point sources by use of a wetland? What is the potential of this wetland for removal of nutrients? Could the non-point sources be completely controlled ? Would it be necessary to enlarge the wetland? How much?

An experimental unit was built in 1984 in the reedswamp, dominated by *Phragmites australis*, to answer these questions. It consists of 12 flow-through basins 1m x 10m x 0.75 m (see Figure 8-1). The basins were designed with no bottom to make as few changes of the flow pattern through the root zone as possible and they were operated with four different hydraulic loadings. The inflow and outflow of water were measured and the precipitation and evaporation were obtained from meteorological data. The exchange of water through the bottom was found by use of water balance calculations. Hydraulic conductivity was different in the basins, however, causing a different exchange of water through the bottom of each basin.

The experiments revealed a clear relationship between hydraulic load and the nitrate reduction in the surface water (Table 8-1). Lower hydraulic loads led to higher nitrate reduction, but higher hydraulic loads led to more nitrate being denitrified per unit of time and area. It is remarkable that, at the highest hydraulic load, the nitrate reduction is as high as 2,700 kg N/ha-yr. The phosphorus retention did not show any relationship with the hydraulic load. Concentrations of phosphorus in inflow and outflow water were approximately the same, although a minor retention was observed during the spring.

The bottom water showed very low nitrate concentrations throughout the year, which implies that the denitrification is almost complete in the depth of 25 cm or more (Table 8-2). The retention of ammonium in the interstitial water varies from 0 to 50 percent, while the phosphorus concentration in the interstitial water was up to ten times higher than in the surface water. Analysis of the interstitial water in deeper, more calcium-rich, layers indicate, however, that the phosphorus washout in the root zone will be re-adsorped there. The release of phosphorus could be explained by the mineralization of organic matter, including phosphorus compounds, by the denitrification process.

The most important processes, namely nitrification, denitrification, mineralization and phosphorus adsorption, were examined in detail in the laboratory to obtain good estimation of process equations and parameters in the model. Table 8-3 shows the results of the examination of the denitrification process. Based upon these investigations, it was possible to estimate that a 0.5 ha wetland—a surprising small area—was sufficient to denitrify all the nitrate in the inflow to Lake Glumsø. This demonstrates the enormous denitrification potential that wetlands possess.

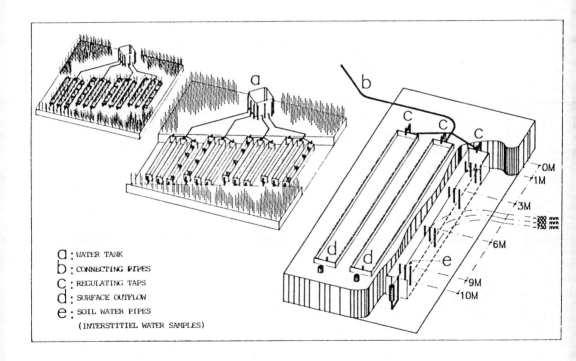

Figure 8-1. Experimental wetland basins placed in a reedswamp at the inlet to Lake Glumsø, Zealand, Denmark.

Table 8-1. Relationship between nitrate reduction and hydraulic load in reedswamp experimental basins, Lake Glumsø, Denmark. One year is set at 270 days due to ice cover during winter. Data are average ± standard error.

Hydraulic Load, liter/m²-day	Nitrate Reduction, percent	Nitrate Reduction, kg N/ha-yr
46 ± 1[a]	65 ± 4	521 ± 17
57 ± 4[b]	65 ± 28	811 ± 168
83 ± 4[a]	62 ± 2	976 ± 6
92 ± 16[b]	71 ± 12	1,077 ± 301
151 ± 18[c]	52 ± 19	1,310 ± 451
312 ± 31[c]	54 ± 26	2,727 ± 1,136

[a] tested in 1984 (n = 3)
[b] tested in 1985-86 (n = 6)
[c] tested in 1984-86 (n = 9)

Table 8-2. Nitrate concentrations in µg-N/l in inlet water and as a function of depth for twelve experimental reedswamp basins at Lake Glumsø, Denmark.

Depth, cm			Inlet Water	
25	50	75	Min	Max
22 - 36	9 - 16	6 - 11	4,000	12,000

The Wet Meadow

The investigations were expanded in November 1985 to a more detailed examination of the nutrient budgets of a wet meadow between agricultural uplands and a small rural stream. The site had an area of 20 m x 120 m. Figure 8-2 gives a three-dimensional picture of the meadow with indications of the borderlines between sand and peat in the sediment.

The nitrate concentration in the inflow water ranges from 15 to 30 mg-N/l, while the concentration in the outflow water from the peatlayer is less than 100 µg-N/l. The flow through the sand layer is unfortunately higher than through the peatlayer and, due to the low concentration of organic matter in this layer,

Table 8-3. Denitrification rates as measured at $10^{\circ}C$ in experimental reedswamp basin receiving 312 l/m²-day (n = 12; 95% confidence limits shown).

Depth, cm	Denitrification, μg-N/g dry wt-day	Dry Weight, percent of wet wt.	Loss on Ignition, percent
0 - 5	273 ± 10	13.8 ± 2.0	50.5 ± 9.4
5 - 10	146 ± 10	13.3 ± 2.1	49.2 ± 8.2
10 - 15	97 ± 10	13.0 ± 1.5	44.1 ± 8.7
15 - 20	56 ± 10	16.7 ± 1.1	32.0 ± 7.1

almost no denitrification takes place there. The flow of water through the wet meadow is low, but there are possibilities to increase it significantly by irrigation. As this is being examined, one of the main objectives of a modelling effort is to be able to predict a total denitrification rate as a function of the hydraulic load.

Presentation of the Applied Model

The model focuses on the possibilities of using wetlands as nutrients traps and to answer management questions such as: How much nitrogen can be removed by denitrification? How much nitrogen and phosphorus is lost to groundwater and neighboring surface waters? How much nitrogen and phosphorus can be removed by harvest of plants? What influence does the hydrology of the wetland have on the nutrient budget? Is it, for example, possible to regulate the hydrology to achieve a more advantageous nutrient budget?

The model is presented in the conceptual diagrams shown in Figures 8-3 through 8-5. The boxes indicate the state variables and the lines indicate processes or pathways. There are 14 state variables in the nitrogen diagram, 11 in the phosphorus diagram, and 5 in the hydrological submodel.

Forcing Functions

The forcing functions of the model are:
- precipitation (process 29 in the hydrologic submodel),
- nitrogen in rainwater (multiplied by 29 gives 6 in the nitrogen submodel),
- phosphorus in rainwater (multiplied by 29 gives 14 in the phosphorus submodel),
- inflows of water (18 to 21 in the hydrological submodel),
- evapotranspiration (30 to 31 in the hydrological submodel),
- temperature in each zone (measured but also found from a relation between soil temperature and air temperature),

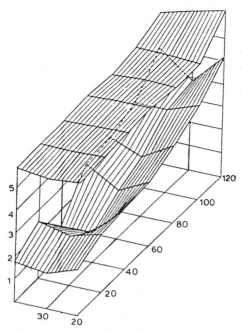

Figure 8-2. Three dimensional profile of Danish wet meadow, showing boundary between peat (above) and sand (below). Numbers are in meters.

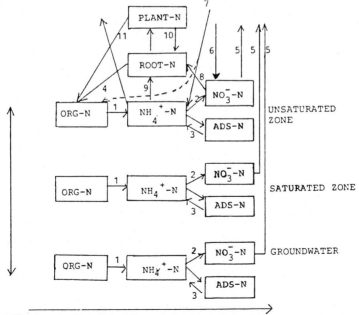

Figure 8-3. Nitrogen submodel of Danish wet meadow. Horizontal arrows flow in one direction, vertical arrows flow upwards and downwards. Some flows are obtained from the hydrologic submodel. Process numbers are identified in text.

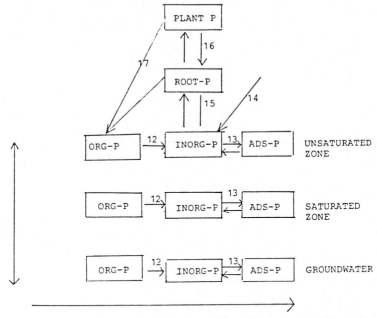

Figure 8-4. Phosphorus submodel of Danish wet meadow. Flows are similar to those in the nitrogen submodel, Figure 8-3.

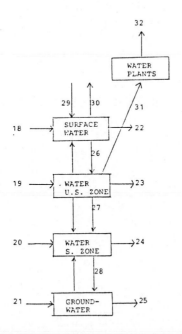

Figure 8-5. Hydrologic submodel of Danish wet meadow. Flows 18 to 21 are horizontal inflows to parcel; 22 to 25 are outflows from parcel; 26-28 are exchanges of water between zones; 29 is precipitation; 30 is evaporation from surface; 31 is water uptake by plants; 32 is transpiration through plants.

- air temperature,
- soil pH, and
- plant biomass.

This last forcing function, plant biomass, is measured as function of time, but for other wetlands it might be found by use of an empirical model, which gives the biomass as function of time for a selected number of various climatic conditions.

Mineralization

The model uses differential equations to express the changes in the state variables. Mineralization, process 1 in Figure 8-3, is described by use of a first order reaction scheme, where the rate constant K is dependent on the temperature T and the moisture content Ω:

$$K = K_{max}\ f(T)\ f(\Omega), \tag{8-1}$$

where $f(T) = K_T^{(T-20)}$ (Bowie et al., 1985), K_T is 1.02 - 1.08, averaging 1.05, while $f(\Omega)$, in accordance with Hansen and Aslyng (1984), is proportional to Ω up to the limit Ω_L and that $f(\Omega)$ decreases linearly between Ω_L and Ω_S, where Ω_S is the moisture content at saturation. It means that:

$$f(\Omega) = \Omega/\Omega_L \qquad \text{for } \Omega \le \Omega_L \tag{8-2}$$

and,

$$f(\Omega) = 1 - (\Omega - \Omega_L)/(\Omega_S - \Omega_L) \quad \text{for } \Omega > \Omega_S \tag{8-2a}$$

Beek and Frissel (1973) presented a table (Table 8-4) for this relationship. The mineralization rate is overestimated by Equation 8-2 at low water content and underestimated at high moisture content. The values in Table 8-4 are used in this model. The mineralization rate is also dependent on the C/N ratio. However, it was considered unimportant in the model at this stage.

Nitrification

The nitrification process, process 2 in Figure 8-3, has also been described as a first order reaction (see, for example, Jørgensen 1983). The influence of temperature and moisture on the rate is expressed in the same way as with the mineralization process, and only the table function relating the rate constant with the moisture content is different (Table 8-5).

Table 8-4. The relationship between moisture content and mineralization rate.

Moisture Content, Ω, % of saturation	Mineralization Function, $f(\Omega)$
0	0
10	0.02
30	0.32
40	0.65
50	0.81
60	1.00
70	0.80
80	0.55
90	0.48
100	0.45

Table 8-5. The relationship between moisture content and nitrification rate.

Moisture Content, Ω, % of saturation	Nitrification Function, $f_2(\Omega)$
10	0.11
30	0.31
50	0.86
60	1.00
70	0.40
80	0.10
100	0.00

Adsorption and Volatilization

The adsorption process (process 3 in Figure 8-3)) is formulated by use of a Langmuir expression, where the constants are found be adsorption investigation carried out on the actual soil. Volatilization (process 4) can be considered a first order reaction and the rate is dependent on the moisture content, pH and the temperature. The dependence of the moisture is simply proportional to the moisture content:

$$f(\Omega) = \Omega / \Omega_S \tag{8-3}$$

The dependence on pH is expressed by use of the following equations:

$$f(pH) = F \tag{8-4}$$

where,

$$F/(F - 1) = 10^{(pH - pK)} \tag{8-4a}$$

and, pK is about 9.2 for ammonium.

As vapor pressure is an exponential function of the temperature, the dependence on temperature might be expressed by use of the following equation:

$$f(T) = e^{(T - 10)/10} \tag{8-5}$$

Denitrification

Process 5 in Figure 8-3 is denitrification and will only be realized under anaerobic conditions, which most probably will correspond to a high water content (closed to saturation). The denitrification rate, r, is often expressed in the literature by use of a Michaelis Menten equation:

$$r = K_d [NO_3\text{-}N] / (K_{dm} + [NO_3\text{-}N]), \tag{8-6}$$

where K_d is a rate constant which is temperature dependent (an equation such as equation 1 could be used) and K_{dm} is the so-called half saturation constant. $[NO_3\text{-}N]$ is the nitrate concentration in g/m^3. Process number 7 is nitrogen fixation, which is not considered in the model at this stage.

Plant Uptake and Translocation

The uptake of nitrogen, U, by the root system of the wetland plants is dependent on the following factors:
    nitrate concentration in the rootzone

- ammonium concentration in the rootzone
- biomass of the roots (BIR), g/m^2
- temperature
- moisture content of the soil
- nitrogen concentration in the roots, (RN), in g per g.

All these factors are considered in the following equation:

$$U = BIR *K_9 \left[[NO_3\text{-}N]/(K_{9m} + [NO_3\text{-}N])\right] \times$$
$$\left[(RNmax - RN)/(RNmax - RNmin)\right] \qquad (8\text{-}7)$$

where K_9 is dependent on the temperature by a similar expression as in equation 1 and by the moisture content by a table function. RNmax and RNmin indicate the upper and lower limits for the nitrogen concentration in the roots. The amount of nitrogen in the roots, ROOT-N, expressed as g per m^2 can be found from the following expression:

$$ROOT\text{-}N = BIR * RN \qquad (8\text{-}8)$$

As mentioned above, BIR as function of time is modelled by an empirical expression, which relates the climatic forcing function and the biomass for the considered plant species. It might also be possible to express BIR as function of time with a constant growth per day up to a certain maximum value for BIR.

Process 10 is the transfer of nitrogen, T, from the roots to the plants. This process is modelled by an equation similar to Equation 8-7:

$$T = PL*K_{10} \left[(RN - RNmin)/(RNmax - RNmin)\right] \times$$
$$\left[(PNmax - PN)/(PNmax - PNmin)\right] \qquad (8\text{-}9)$$

where PL is the amount of plant biomass per m^2 and PN is g/g of nitrogen in plants. PNmax and PNmin indicate the upper and lower limits of nitrogen in the plants. The limits are dependent on the time and might be given as tables or graphs as in Figure 8-6. The nitrogen content in plants (PLN), expressed as nitrogen per m^2, is found as:

$$PLN = PL * PN \qquad (8\text{-}10)$$

Process number 11 is the transfer of dead organic matter from plants and roots to organic nitrogen. It takes place over a period from autumn to the beginning of spring the following year. This process is described as a temperature dependent first order reaction with an almost complete transfer during the indicated period of time.

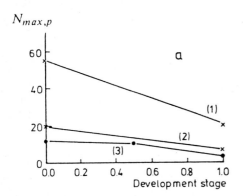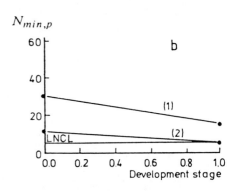

Figure 8-6. The maximum (N_{max},P) and minimum (N_{min},P) nitrogen concentration for leaves (1), non-leaf material (2), and roots (3) of natural grassland vegetation as a function of phenological age (development stage, here defined as being 1.00 at maturity) (Jørgensen, 1986).

Phosphorus Processes.

Process number 12 is the mineralization of the organic phosphorus components, which is described in a similar way as the mineralization of organic nitrogen, only the rate constant might be different. Similarly, processes 13-17 for phosphorus are expressed with the same type of equations as for the parallel nitrogen processes, but the parameters might be different.

Hydrologic Submodel

The hydrological submodel is shown in Figure 8-5. It accounts for the flows of water and describes therefore the transport of solutes from one parcel to another (horizontal flows) and from one zone to another (vertical flows). Four layers or zones are considered: the surface layer (5 cm), the root zone or unsaturated zone minus the surface layer, 45 cm, the next 50 cm, and the lowest 100 cm.

The processes 18-21 are inflows of water to the four layers. They are forcing functions to be determined from empirical measurements. The vertical and horizontal flows 22-28 are given by Darcy's equation:

$$v = K_H \, (\pi_1 - \pi_2) / l, \qquad (8\text{-}11)$$

where v is the flow rate, K_H the hydraulic conductivity, π_1 the potential at point 1 and π_2 the potential at point 2. The distance between the two points is l. K_H and π are functions of the moisture content ß. These functions will be found through empirical measurements.

The wet meadow was divided into 12 parcels and as each parcel has 4 zones or layers, 48 geometric boxes are considered. The model accounts for the amount of water, Q, in each box and the numbers 1-48 are used to indicate which box we consider.

The hydraulic capacity is determined for each box and the moisture content ß is simply determined by the following expression:

$$\text{ß}_n = Q_n / HC_n \qquad\qquad (8\text{-}12)$$

where HC is the hydraulic capacity. HC is not a constant but increases with ß (ß increases as more and more pores are filled). When Q = HC, further inflow to the box will stop.

The processes 29-32 are forcing functions. Process 29 is precipitation and processes 30-32 are the evapotranspiration. The total result of these three processes are determined, but they are shown as three processes, because it might be convenient on a later stage to consider all three processes independently.

Each of the flows 18-28 is accompanied by a flow of solutes including ammonium, nitrate and inorganic dissolved phosphorus. In addition, diffusion takes place, but it was found negligible compared with the advection.

State variables and forcing functions are shown in Tables 8-6 and 8-7. All differential equations are simply formulated as follows:

$$dX/dt = \Sigma \text{ inputs} - \Sigma \text{ outputs}, \qquad\qquad (8\text{-}13)$$

where X is a state variable. The program is based on the simulation language CSMP. Rather good estimations can be found for the parameters in the literature (see, for example, Clark and Rosewall, 1981; Jørgensen, 1979; and Penning deVries and Van Laar, 1982).

Model Results

The presented model has at this stage not yet been properly calibrated and validated. However, it has been attempted with a preliminary calibration, which can be improved, when more data are available, to simulate various management strategies and to examine which factors are most important for the function of the wetland as nutrient trap. These results are summarized below:

1) The denitrification rate is independent of the nitrate concentration, provided it is at least several mg/l.

2) If the concentration of organic matter is high (peat), the denitrification potential under anaerobic conditions (which is found a few cm under the surface) is about 14 g NO_3-N/m^3-day at 20°C. If the concentration of easily

Table 8-6. State variables for wet meadow model.

Name	Units	Total Number of Variables
organic N	g N/m^3 soil	48
soil moisture	ratio: water/capacity	48
NH$_4$-N	g N/m^3 soil	48
NO$_3$-N	g N/m^3	48
Root N	g N/m^2	12
RN	g N/kg	12
PLN	g N/m^2	12
PN	g N/kg	12
ADS-N	g N/m^3	48
organic P	g P/m^3 soil	48
INP	g P/m^3 soil	48
ADS P	g P/m^3 soil	48
Root P	g P/m^2	12
PLP	g P/m^2	12
K$_H$	m/day	48
π	(potential) m^{-1}	48
Q	m^3	48

Table 8-7. Forcing functions for wet meadow model.

Name	Unit	Source
PREC	m/day	KVL meteorological station
N$_p$	g N/m^3	literature/measurements
P$_p$	g P/m^3	" "
Inflow Water	m^3/day	measurements
Evapotranspiration	m/day	KVL meteorological station
Aerobic/Anaerobic	---	measurement and/or f(ϑ)
Temperature	$^{\circ}$C	measurements of function of air temperature
Solar radiation	energy/m^2	KVL meteorological station
pH of soil	---	measurements
root biomass BIR	kg/m^2	"
plant biomass PL	kg/m^2	"

biodegradable organic matter becomes limiting, the above-mentioned potential is proportional to the concentration of organic matter.

3) The hydraulic load should be adjusted to this denitrification potential to optimize the use of the denitrification potential.

4) The soil has a certain adsorption capacity to phosphorous compounds. This capacity is, however, limiting and harvest of plants seems to be the only long-term method to use wetlands for phosphorus removal.

5) It is possible, through the proper management of wetlands, to remove a considerable amount of nutrients, particularly nitrate nitrogen.

Conclusions and Recommendations for Future Modelling Efforts

The experiences gained by these wetland studies, which are still not completed, demonstrate that it is indeed possible to construct models of the nutrients cycles in wetlands. The studies have further shown that wetlands may be very important tools for solving the problems of the non-point sources. Because modelling of wetlands, particularly for systems such as these wet meadows, is early in its development, there are several recommendations and conclusions that follow.

Use a Hierarchy of Models

A series of models should be considered for the study of wetlands as nutrient sinks, ranging from site-specific detailed hydrologic models to a general Vollenweider-type empirical model for wetlands as interface systems for controlling agricultural runoff. Modelling of ecosystems need not rely on only one modelling approach or one model to maximize its usefulness. In fact, it is appropriate in the beginning of an ecological investigation to develop a series of models in an hierarchical or non-hierarchical framework. Table 8-8 suggests four modelling approaches that might be appropriate for our wet meadow wetland. The first is a detailed hydrologic model that could be used to organize present hydrologic data collection efforts and could be used to verify hydrologic measurements (and vice-versa). This model could most appropriately be an "off the shelf" subsurface model, many of which have been developed or it could be further development of the hydrologic submodel described above and in Figure 8-5. An ecosystem model, such as the one described in this paper, has less hydrologic detail but more ecological detail, shows a quasi-spatial view of the wetland site and would give details of the biogeochemistry. A third model, which could be developed later when the experimental data and previous two models are nearing completion, could consider the spatial aspects of a river basin where these meadows are developed along the shoreline of the river for some length. This model could be a combination of a stream model with simpler ecosystem model of the meadow wetlands. This model could lend itself to development of novel approaches to displaying spatial results (see, for example, Sklar et al., 1985 and Chapter 6).

Table 8-8. Types of models considered for Danish wet meadow study.

Model Name	Purpose of Model	Spatial Scale	Time Scale	Software/ Computer
Hydrologic	to give details of hydrologic flow	100 m x 1 m x 1 m	days/ weeks	hydrologic model on mainframe
Meadow Ecosystem	to demonstrate nutrient cycling and processes	1 m^2 x 48	1-3 years	microcomputer/CSMP
Spatial Stream/ Meadow model	to show spatial patterns along streams	100 m x 10 km	days/ weeks	transport/ mainframe
Empirical Loading Model	estimate amount of meadow needed for given flows	none	static	simple graphs/ calculator

Empirical Wetland Models Should be Continued

The fourth model is not a simulation model, but a management-oriented empirical model similar in function to the Vollenweider nutrient loading model (Vollenweider, 1968) but for matching agricultural nutrient inputs with the appropriate size and type of riparian wetland ecosystems. The two studies referred to here are based on an extensive data collection. It is obviously not possible to gather extensive data for each wetland to be modelled. Therefore it should be possible to construct a simpler model to be used more generally. The idea can be expressed as follows: given the profile of easily biodegradable organic matter for the wetland, the possible hydraulic loads, the nitrate concentration of the inflowing water, the areas and the main species of vegetation the wetlands, and the model should be able to give an estimation of the amount of nitrate which can be denitrified and the amount of phosphorus and nitrogen which can be removed by harvest. The two studies have made it possible to select the most important variables in such a simple and general model.

Models Should be Useful for Wetland Managers

Consideration needs to be given to the ultimate use of a model in a management situation. Detailed spatial modelling of the wetland is probably not necessary to answer the nutrient retention questions being posed for many wetland systems. The spatial pattern that might be of most interest to water resource managers is longitudinally along the streams and bodies of water that are to benefit from the nutrient removal capabilities of the wetland. While a spatial model of the specific wetlands could be of benefit in verifying the model, this scale does not aid the decision maker in determining whether or not wetlands are effective nutrient sinks. Figure 8-7 indicates that the empirical loading model might be the most "useful" to decision makers for such situations. This does not preclude the use of the other models to verify parameters or explore eco-system theories.

Emphasize Wetland Intrasystem Processes

There is a need to give attention to the above-ground wetland ecosystem processes as well as the subsurface processes. This was recognized in the above model, with the vegetation shown as a seasonal forcing function. Surface vegetation (and the accompanying root system) may play a very important role in the processing of nutrients in certain wetlands. At a minimum, good estimates

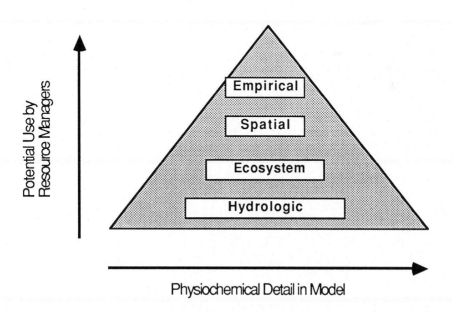

Figure 8-7. Estimate of usefulness of various wetland models to wetland managers.

should be obtained on the growth rates of the dominant surface vegetation. Estimates of gross primary productivity and respiration would be best, but net primary productivity should be estimated on a seasonal basis at a minimum. Decomposition studies will aid in determining the rate at which nutrients are recycled to the soil. Periodic measures of nutrient content of the detritus should be made to investigate changes in nutrient concentrations. The hydrologic regime of these meadows will probably be changed considerably with the addition of agricultural surface runoff, so the effects of hydrology on vegetation productivity should be considered as well.

Develop Preliminary Nutrient Budgets

Preliminary nutrient budgets, showing approximate flows on a seasonal, monthly, or yearly basis, should be developed immediately for any wetland system where nutrient retention is being considered. This will give the modellers some preliminary data to develop the modelling strategy and will provide a preliminary review of data collection efforts. This is particularly important so that the most significant processes and forcing functions can be given more attention in the data gathering phase. These budgets can be updated as more data accumulates.

Consider Connections with Adjacent Ecosystems

Significant attention needs to be paid to the exchanges between a wetland and its adjacent aquatic system. In our study sites, the reedswamp experimental wetland site flowed into a lake, while the wet meadow discharged into a small stream. The exchanges between the meadow and the stream primarily involve the yearly slow input of groundwater from the meadow to the stream, but a significant winter flooding of the meadow by the stream also occurs. Sediments will accompany the flooding and are often rich in nutrients, thereby "reversing" the normal flow of nutrients to the stream. These deposited nutrients could lead to an even greater nutrient retention by the wetland than calculated by a simple budget.

Hydrologic Measurements are Paramount

Hydrology is the most important control on wetland functioning (Mitsch and Gosselink, 1986) and a sound hydrologic budget needs to be developed for any modelling effort. When groundwater flows are involved, this becomes even more important. Vertical and horizontal permeabilities (hydrologic conductivity) are often the most important hydrologic variables when groundwater is involved and yet these coefficients can vary over many orders of magnitude in one wetland, so careful selection of sampling sites should be included in their measurement. Evapotranspiration remains a major yet poorly understood hydrologic flow in wetlands. The general question as to whether wetlands pump more or less water than open bodies of water is not answered yet, nor is the answer probably the same for all sites. This is important because subsurface flow is often estimated as the residual of inflows minus evapotranspiration. Interception and subsequent evaporation of water on the stems of the plants should also be considered in the hydrologic budget. Not all of the water reaches the soil surface in wetlands, especially in low intensity rainfalls.

References

Beek, J. and M. J. Frissel. 1973. *Simulation of Nitrogen Behavior in Soils.* Center for Agricultural Publishing and Documentation. Wageningen, The Netherlands, 67 pages.

Bowie, G.L. et al. 1985. *Rates, constants, and kinetics formulations in surface water quality modeling.* U. S. Environmental Protection Agency, Environmental Research Laboratory, Athens, Georgia.

Clark, F.E. and T. Roswall, editors. 1981. *Terrestrial nitrogen cycles: processes, ecosystem strategies and management impacts.* Ecol. Bull. (Stockholm) No. 3, 714 pages

Hansen, S. and H. C. Aslyng. 1984. *Nitrogen balance in crop production. Simulation model NITCROS.* Hydrotechnical Laboratory. The Royal Veterinary and Agricultural University, Copenhagen, Denmark.

Jørgensen, S.E., editor. 1979. *Handbook of Environmental Data and Ecological Parameters.* International Society for Ecological Modelling, Copenhagen, Denmark, 1162 pages.

Jørgensen, S.E., editor 1983. *Application of Ecological Modelling in Environmental Management, Part A.* Elsevier, Amsterdam, 735 pages.

Jørgensen, S.E. 1986. *Fundamentals of Ecological Modelling.* Elsevier, Amsterdam, 389 pages.

Mitsch, W.J. and J.G. Gosselink. 1986. *Wetlands.* Van Nostrand Reinhold, New York, 539 pages.

Orlob, G. T., editor. 1983. *Mathematical Modeling of Water Quality: Streams, Lakes, and Reservoirs.* John Wiley and Sons, New York, 518 pages.

Penning deVries, F.W.T. and van Laar, H.H., editors. 1982. *Simulation of Plant Growth and Crop Production.* Centre for Agriculture Publishing and Documentation, Wageningen, The Netherlands. 308 pages.

Sklar, F. H., R. Costanza, and J. W. Day, Jr. 1985. Dynamic spatial simulation modeling of coastal wetland habitat succession. *Ecol. Modelling* 29: 261-281.

Vollenweider, R.A. 1968. *Scientific fundamentals of the eutrophication of lakes and flowing waters, with particular reference to nitrogen and phosphorus as factors in eutrophication.* Rep. Organization for Economic Cooperation and Development, Paris, 192 pages.

9/ SOME SIMULATION MODELS FOR WATER QUALITY MANAGEMENT OF SHALLOW LAKES AND RESERVOIRS AND A CONTRIBUTION TO ECOSYSTEM THEORY

M. Straškraba
P. Mauersberger

Two opposing approaches to model complexity exemplified by models developed for shallow flow-through lakes in the German Democratic Republic (GDR) and for reservoirs in Czechoslovakia are demonstrated: detailed realistic models with many variables and simple but more general models, respectively. The advantages and disadvantages of the two directions with respect to solving water quality problems and to theoretical predictions are specified and future modelling directions outlined. To overcome shortcomings of the present modelling methodology, two approaches, thermodynamic and cybernetic (control theoretic), are presented. In addition, an attempt to combine both for more profound theoretical understanding of aquatic ecosystem processes is included.

Introduction

In wetland areas, open waters often occupy an important fraction of the total area. This fraction can be further increased by construction of dams, resulting in artificial lakes—reservoirs. The purposes of construction can be rather varied: water-supply, irrigation, flood-protection, fishery and power generation, and often several purposes are combined. The variety of models used and of problems tackled is considerable.

In this contribution we focus on modelling activities of two academic laboratories in GDR and Czechoslovakia (see authors' affiliations) which have two common interests: application of models to water-quality problems and focus on novel methodology. In water quality modelling, two opposite directions with respect to model complexity can be recognized: to build as complex and detailed specific models as data and personnel permit, or to construct relatively simple but more general models. The two laboratories represented in this paper use different modelling approaches: *The Department of Hydrology, Institute of Geography and Geoecology (GDR)* uses complex models, while *The Hydrobiological Laboratory, Institute of Landscape Ecology (Czechoslovakia)* employs simple ones. With respect to this, it would perhaps not be without interest to compare the two approaches, in spite of the fact that different water body types are considered: shallow lakes on one hand and reservoirs on the other (Table 9-1). Identity is in scope — both modelling activities are related to water quality problems. In addition, both authors are dissatisfied with the present methodology of ecological simulation modelling and are investigating new directions with possible reflections on the development of the theory of ecosystems and of their management. Originally, two approaches toward this goal were developed independently, one based on thermodynamics and the other on cybernetics (control theory). Recently, attempts to compare and possibly combine the two directions to reach a common goal were made. In this paper we attempt to present, in addition to a review of the results obtained up to now, some outline of the underlying philosophy.

Aquatic Ecosystems and Their Surroundings

Aquatic ecosystems are open systems closely coupled energetically, materially and informationally to their surroundings (Figure 9-1). For throughflowing lakes and reservoirs, the most important coupling is by in- and outflow. The consequences of throughflow for reservoir limnology are rather extensive and concern all aspects, starting from their hydrophysical through hydrochemical up to biological processes. Based on empirical and simulation studies of several reservoirs in Czechoslovakia, we consider hydraulic retention times a key factor of reservoir limnology and similar consequences can be drawn from studies of shallow lakes in GDR. Uhlmann (1972) calls reservoirs "reactors" comparable to a throughflowing continuous cultivation of organisms and this is actually the representation our dynamic simulation models for all water bodies take.

The consequence of this open character is that processes of water quality in a reservoir are closely coupled with those in the watershed, and that they considerably affect the outflowing river. From the water quality point of view, the most important components of the inflowing water are the organic matter concentration (BOD, COD) and nutrient loading. The study by Straškrabová (1975) demonstrates that an effect of a reservoir on the outflowing water quality can be highly positive up to considerably negative, depending on inflow pollution on one hand and organic production in the reservoir on the other. The governing parameters have being quantified into an empirical mathematical model.

Table 9-1. Comparison of the eutrophication models developed in the Institute for Geography and Geoecology (German Democratic Republic) and Institute of Landscape Ecology (Czechoslovakia) respectively.

	IGG (GDR)	ILE (Czech.)
State variables	7 to 22	3 (5)
Feedback mechanisms	few	many
Morphology	site specific	generalized, lake size dependent
Circulation	full mixing (shallow)	stratified (deep)
Limiting nutrient	nitrogen, phosphorus	phosphorus
Effects of fish	not considered	considered
Processes in sediment	described in detail	not included (but see AQUAMOD 3)
Applied to bodies of water	Muggelsse, Lakes of the Havel-System	Talsperren, Slapy, Klíčava, Orlík, Rímov, Želivka
Theoretical development	thermodynamic considerations	cybernetic (control theoretical)
Practical application	drinking water supply from lakes	eutrophication management of reservoirs

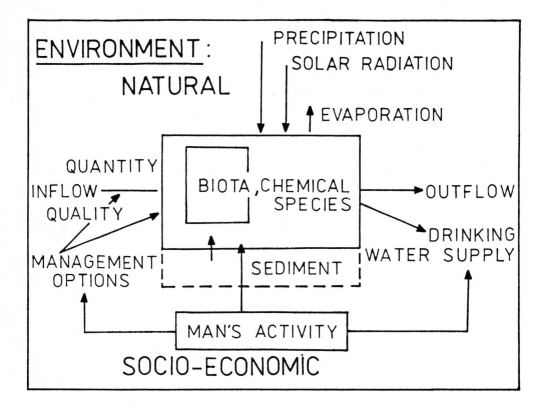

Figure 9-1. Aquatic ecosystem model showing an open system coupled to its surroundings.

As to nutrients, a striking feature in developed countries is the trend of increasing loading of both nitrogen and phosphorus. When such trends started to be recognized, their character seemed to be linear—some constant fraction of nutrients added by man's activity to watersheds appeared to be leaking into water bodies. Recent evidence suggests that concentrations seem to rise exponentially rather than linearly, which is becoming even more dangerous than before (Procházková, 1983). It indicates that some kind of saturation level of the watersheds by nutrients was reached and an increasing proportion is released to water.

Modelling of Lake and Reservoir Hydrodynamics

Hydrodynamics and thermics of a standing water body represent a matrix inside which all other processes are played on. For this reason Nihoul (1975) coined it "support of the ecosystem." For the particular ecosystem types (throughflowing lakes and reservoirs), we are faced with the problem of how thorough the representation of this matrix should be. On one hand, we cannot hope to understand the chemical and biological processes if we do not have enough understanding for hydrodynamic variability as affected by hydrometeorology, throughflow regulations, outlet manipulation etc. On the other hand, there are a number of urgent problems of chemistry and biology, and their coupling with physics does not directly follow only from hydrodynamic modelling.

In regard to these essentially manpower restrictions, two directions of representing variable hydrodynamics were followed in our laboratories: an empirical (original simplified description based on statistical summarization of extensive data sets) and an analytical (largely based on simulation models developed in other laboratories).

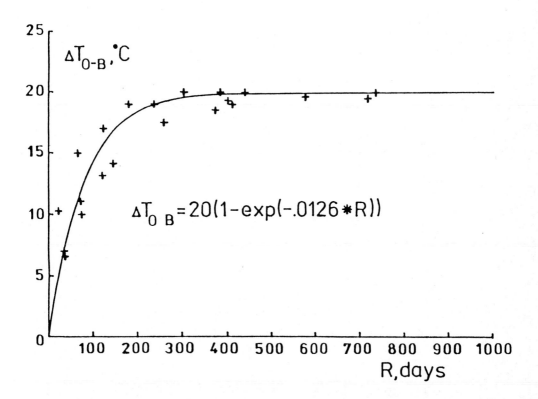

Figure 9-2. Plot of the simplest measure of reservoir stratification, the summer surface–deep water temperature difference (ΔT_{O-B}) against the theoretical retention time (R) of the reservoir.

The first direction is used here only to demonstrate some systematic differences in stratification of deep valley reservoirs with hypolimnetic outlets. Figure 9-2 is a plot of simplest measure of stratification, the summer surface–deep water temperature difference against hydraulic retention time. Here the temperature of 30 m is used as a reference frame for deep water. The figure demonstrates that for reservoirs in a similar geographic location there is some critical retention time at which full stratification is observed, whereas below that level stratification is disturbed by the throughflow.

The idea of the dependence of temperature stratification in reservoirs on retention time is elaborated quantitatively in the model RESTEMP (Straškraba and Gnauck, 1985). The model is based on elaboration of annual observations of temperature at several depths in a number of reservoirs in Czechoslovakia and GDR with different hydraulic retention time. For each depth the annual course of temperature was approximated by means of periodic regression according to the equation:

$$T(zt) = T_z + \sum_{i=1}^{2} A_{zi} \sin(i \cdot t + \gamma_{zi}) \pm e \qquad (9\text{-}1)$$

For shallower depths, some 90-95% of total variance is given to the first harmonics, as shown earlier for lakes (Straškraba, 1980a). The participation of the second harmonics increases with depth due to slower warming of the deeper strata in spring and summer but rather rapid cooling during autumnal overturn. In the model, however, only the first harmonics, i.e., the coefficients $k_i = T$, A_1 and γ_1 (T = average annual temperature, A_1 = semiamplitude of annual variation, γ_1 = phase shift) are taken into consideration. For each observation year, the depth distribution of the parameters k_i ($k_i = f(z)$) was approximated by means of an exponential function:

$$k_i = k_{i,0} e^{-b_i z} \qquad (9\text{-}2)$$

The coefficient b_i represents the rate of decrease of the respective parameter with depth. Larger b_i means a steeper drop of T, A_1 or γ_1 with depth. In the model RESTEMP coefficients b_i are related to theoretical hydraulic retention time (R, days). The model shows that, for R < 100 days, stratification deflects markedly from the "lake" figure, surface summer temperatures being lower and deepwater temperatures higher than would be predicted by the lake model. Additional reduction to R = 50 days causes an increased throughflow which mixes the strata more rigorously, and stratification is decreased. At R = 10, the water body is no longer stratified and conditions approximate those of a very deep river.

Hydrodynamic simulation models which were taken as a starting point for our purposes are different according to water body. Hydrodynamics of shallow lakes seems to be dominantly driven by wind mixing and a water-flow model seems appropriate. As shown above, for reservoirs the dominant driving force is hydraulic retention time. One of the models elaborated until now seems particularly suited to our goals: Markofsky and Harleman (1971). Based on an empirical thermal budget coupled with mass-balance, temperatures and exchange rates between individual layers are calculated. Daily meteorological values, flows and surface elevations are inputs to the model, the shape of the reservoir being also specified. The model seems to be one of the most extensively applied models, both for lakes (Imboden et al., 1983) and reservoirs (Henderson-Sellers and Archer, 1982). Our application to temperature stratification of Klicava Reservoir during 1967 has shown a good agreement with observations for this year (Figure 9-3).

However, when applying the model to other reservoirs in Czechoslovakia (Slapy, Orlík, Želivka, Rímov and Domaša) we recognized the following shortcomings:

1) The model appeared very sensitive to values of the extinction coefficient for visible white light (ETA) (Figure 9-3). In the model, however, ETA is given as constant throughout the year, and the observed annual variations of ETA are not considered by the model.

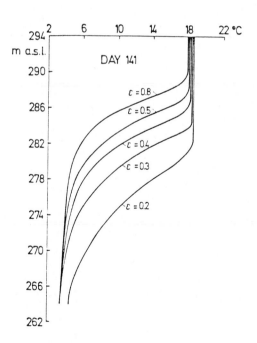

Figure 9-3. Depth profiles of temperature calculated by means of the hydrodynamic model. The situation corresponds to Klíčava Reservoir 1967, day 141 (mid-August). Different values of E (= ETA) are given, suggesting the sensitivity of the model to this parameter, the extinction coefficient for visible white light.

2) The model is rather sensitive to solar radiation inputs, which are usually not directly measured at all corresponding sites. Cloudiness is measured much more commonly.

3) The processes of intensive mixing of water masses in reservoirs with short retention times (R < 100 days) are not represented in the model. This results in disagreement between observations and data for such conditions.

4) A version of the model including DO calculations we call OXYGEN is inadequate for eutrophic water bodies. It calculates oxygen in a very simplified (Streeter-Phelps) form, i.e., as a first-order decay process. The effects of organic production and its decomposition are not at all respected.

5) The model belongs to the category of two-dimensional ones with dimensions time and depth (spatially one-dimensional model). Horizontal changes are represented only by means of very simplified empirical approximations. It is impossible to respect a complicated shape (e.g. with several major inflows), to model longitudinal and/or transversal differences.

We have tried to overcome some of the shortcomings mentioned. As to point 2, we have formulated a subroutine using results of the model SORA (Straškraba, 1980a). Our subroutine FLXIN calculates solar radiation inputs from sinusoidal approximations of theoretical values for direct and diffuse radiation obtained from SORA for a given geographical latitude. The reduction of radiation due to clouds is calculated from observed cloudiness and an empirical formula in SORA. Presently we are working on a modification of a program for overcoming problems 1 and 3. The input of ETA = f(t), according to observed data presents no difficulty. Problems arise when the effect of algae on ETA has to be considered, because this means modelling algal growth separately. As to the effect of throughflow (retention time) on stratification, the only way to improve model performance seems to be to reflect increased turbulence by means of the shearing stress or to derive the "corresponding" increase of turbulence from the empirical relations discussed earlier.

As to point 4, the version OXYGEN was modified by our colleague Chytil (1980) for calculations of the fate of a radioactive substance coming into the waterbody.

Until now the two hydrodynamic models, for wind-driven circulation in shallow lakes and for inflow-outflow driven mixing in reservoirs, were elaborated independently; we can see here the following possibilities for collaboration:

1) Development of methods of coupling the effects of wind as represented in the shallow lake model with the effect of flows as given by the reservoir model. Similar extensions were already attempted earlier by the originators of the model (Octavio et al., 1977).

2) Development of methods of coupling of the hydrodynamic and chemical-biological models. This is a question which deserves more thorough study because of its broad applicability. It is of course possible to make an ad hoc new code for combining different models. However, this is not very efficient, keeping in mind that future developments of both the hydrodynamic and the ecologi-

cal submodels can be fully independent in relying on the work of different specialists. In addition, we may expect in the future a need to combine more than two models, and a generalized model-building strategy for such situations would be useful.

The problem is relatively straightforward when the submodels, or the processes represented by them, are mutually independent. In this case the computation can be sequential in the sense that output of submodel 1 is used as an input for submodel 2. However, ecological processes are usually interrelated, as is the case for hydrodynamic and biological processes. We have shown above that the model used is very sensitive to ETA, and this is a variable dependent on algal biomass, which itself is affected by thermal (density) stratification as well as directly by temperature. There is also a relationship between chemical concentrations and density, but this is usually negligible for situations most commonly encountered. In the interrelated subsystems case, the submodels should be coupled in a feed-back type mode (Figure 9-4). The problems of unequal time steps and of partial differential equations of unequal size steps have to be solved, particularly when variable step methods are used and automatic stability criteria are involved.

Simulation Models of Eutrophication and Water Quality

EMOS is a macroscopic, deterministic model developed in GDR. It represents the aquatic ecosystem of shallow lakes exchanging matter with the sediment and with neighboring water bodies (Schellenberger et al., 1978, 1983). Seventeen chemical and biological compartments can be taken into account in every element of a cascade of mixed compartments, the number of which is arbitrary. The model represents the energy fluxes in the ecosystem, stoichiometric cycling of the nutrients phosphorus and nitrogen, horizontal flows of chemical and biological components between the segments, exchange of nutrients and organic matter between sediment and pelagial, benthic components and fish.

External chemical and biological loads, solar radiation and water temperature are the driving forces. The following constituents are used: in the pelagic zone—dissolved inorganic nitrogen, dissolved organic nitrogen, dissolved orthophosphate, dissolved nonorthophosphate phosphorus, detritus (in a special version of the model divided into a labile and a refractory component), two or three groups of phytoplankton, zooplankton, fish, heterotrophic bacteria; in the benthic zone—dissolved inorganic nitrogen and phosphorus in the interstitial water, sediment sorbed phosphorus, bottom detritus, phytobenthos, zoobenthos and bottom bacteria.

The model consists of a system of nonlinear ordinary differential equations of the first order. Hydrodynamic transport processes are calculated by a simple submodel. Another submodel describes the sediment-water relationship, taking into account bacterial and autolytic decomposition of the bottom detritus, chemical fixation and adsorption of phosphorus by the sediment,

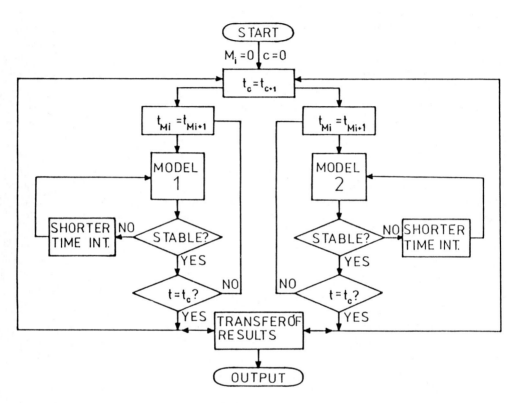

Figure 9-4. Suggested block diagram for coupling of two interrelated simulation models, e.g. the hydrodynamic and biological models.

release of sorbed phosphorus from sediments, and phytobenthic assimilation.

The related model EMSY is comprised of 7 components (2 types of algae, zooplankton, phosphorus, nitrogen, and detritus in pelagial and sediment). The actual wind velocity acts as one of the driving forces controlling biological processes in the pelagial and the exchange of material between water and sediment. The input of hydrological and meteorological data in relatively short time intervals aims at forecasting water quality for the next weeks (Behrendt and Schellenberger, 1986).

As seen, the models developed in the IGG Berlin are rather complex. Basic philosophy used in the Czechoslovak laboratory is rather different in this respect. Some differences given in Table 9-1 are related to the character of localities to be represented by the models. For example, for Czechoslovak reservoirs, phosphorus appeared to be limiting, whereas for northern German lakes and particularly for brackish ones nitrogen is, under certain circumstances, more important than phosphorus. Other differences are connected with the major scope of modelling activity: to represent one or a few lakes by ad hoc models which will allow distinguishing their specific features on one side and, on the other to emphasize modelling aquatic ecosystems and their averaged reactions as observed in nature by comparative studies of many localities and years.

When comparing the two approaches it is not possible to say that one is preferable to another. Each methodology has its advantages and disadvantages. Complex models are too complex to be completely understood, verified and utilized, and simple models are too simple to represent reality with sufficient accuracy. Therefore, some kind of synthesis of the two approaches must be obtained in the future.

As an attempt in this direction, members of both laboratories have coupled the generalized two-layer model AQUAMOD 2 (Dvořáková, 1976) with a simplified version of the phosphorus-exchange sediment model (Kozerski, 1977). As a result, the model AQUAMOD 3 was obtained. The model represents dynamics of the phosphorus budget of a water body with balanced representation of physical, chemical as well as biological processes. In sediment, the decomposition of sedimented organic matter (of algal origin) takes place, with autolysis representing a rapid process and decomposition representing a slow one. Adsorbed and dissolved phosphorus result, adsorption being more rapid during anoxic than at oxygenated conditions. It also depends on the concentration of Fe (or other absorbants) in sediment. Dissolved phosphorus is exchanged with the upperlaying water by turbulent mixing. A part of organic matter is compacted to deeper permanent sediment strata. When recent sedimentation is low, some organic matter and phosphorus may also enter the active layer from deep sediments.

The first testing of the sediment submodel was made by Kozerski and Dvořáková (1982) using the results of earlier laboratory experiments by Tessenow. AQUAMOD 3 was also used for studying some features of phosphorus cycling in lakes. Annual cycles were calculated for different conditions, and simultaneously the rates of all phosphorus exchange processes were followed (Straškraba and Gnauck, 1985). Figure 9-5 is a schematic representation of the phosphorus cycling through free water, algae and zooplankton in the epilimnion and hypolimnion as well as through sediment in different annual phases in a deep mesotrophic lake with fairly high throughflow. We can learn from the study that free phosphate-phosphorus in the epilimnion is but a small fraction of total lake phosphorus, but it is extremely mobile and recycled rather rapidly in dependence on biological activity. The enormous phosphate pool in hypolimnion is much more static, the exchange with the epilimnion depending on both turbulent mixing and epilimnic biological activity. In sediment, a very active process is the exchange of free and adsorbed phosphorus.

Although it is only a highly idealized and simplified picture of reality, the model AQUAMOD 3 represents a means for demonstrating and quantitatively following major features of dynamics of processes in aquatic ecosystems.

Application of Models to Water Management Problems

The concentration of quantitative knowledge about the dynamics of aquatic ecosystems into simulation models opens the possibility of their effective use for solving some practical problems. Answers to some questions concerning prediction, planning and management of shallow lakes and reservoirs can be obtained on the basis of simulation results.

The model EMOS was initially developed, calibrated and verified using data from a chain of shallow inlets along the Baltic Sea coast in the North of the GDR between Rostock and Stralsund (Area: 197 km^2; mean depth: 1.7 m). Four compartments have been used. The input from the drainage basin (9.7 x 10^5 m^3/day) and the influence of the exchange with the brackish water of the Baltic Sea ought to be controlled in order to ensure water quality requirements in the inlets. Annual variations of nutrient concentrations, phytoplankton and zooplankton were simulated. Inter alia, it resulted that a very high percentage of the imported phosphorus is retained in the sediments of the inner inlet regions. The model has also been applied to the aquatic ecosystem of Lake Grosser Muggelsee (Berlin). The river Spree flows through this lake (Area: 7.7 km^2; mean depth: 4.8 m; theoretical retention time: 38 days with an average discharge of 10^6 m^3/day). On the one hand, the model was used for studying the ecosystem behavior. On the other hand, proposals for management strategies have been derived. The model shows the cycling of nutrients through the ecosystem, the strong interrelationship between P-dynamics in the pelagial and the sediment, the role of zooplankton and bacteria for the remineralization of detritus etc. Scenarios simulate the state of the lake when the loadings are assumed to be reduced. Import of particulate organic and inorganic matter prove to influence deeply the bioproductivity and water quality of the lake and, therefore, must be reduced. The models EMOS and EMSY also have been applied to other lakes and to shallow reservoirs in the GDR and USSR in order to get information about optimum management measures. By reducing 75% of the nutrient input, seston is significantly diminished, seston maximum is shifted into late autumn, and anoxic release of phosphorus from the sediments is stopped. However, bluegreens continue to dominate in midsummer, and diatoms are dominant during the remaining months of the season.

In Czechoslovkia, the models are applied particularly to problems of drinking water supply from reservoirs (Straškraba and Straškrabová, 1975; Straskraba 1982). More empirical models based on the BOD changes in reservoirs (Straškrabová, 1975) and on the chlorophyll–phosphorus relationship (Straškraba et al., 1979) have found wide application. Simulation models were applied to water quality development in the reservoirs Slapy and Orlík, to optimal sizing for Slezská Harta Reservoir, to estimation of the effect of water-transfer from Sázava River to Želivka Reservoir and drinking water supply of Prague and to determination of possibilities of improving the water quality of

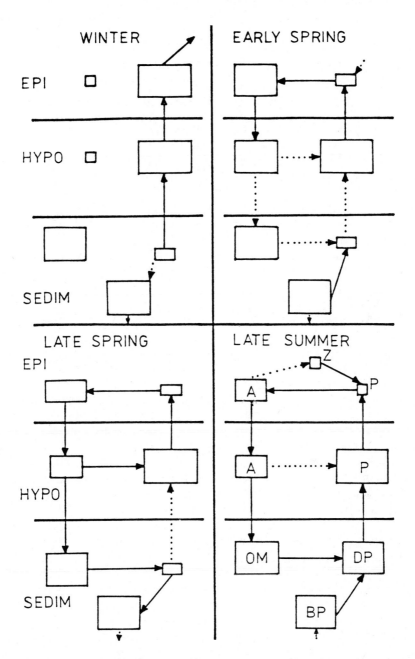

Figure 9-5. Schematic representation of the phosphorus pools and cycling in a stratified mesotrophic lake calculated by the model AQUQMOD 3. The relative size of the pools (see designation in the right lower part) is proportional to box sizes. Major flows are given by heavy lines, less intensive flows by dotted lines. EPI = epilimnion; HYPO = hypolimnion; SEDIM = topmost layer of sediment; A = algae; P = dissolved phosphorus; Z = zooplankton; OM = fresh organic matter in sediment; DP = dissolved phosphorus in the interstitial water; BP = bound phosphorus in the interstitial water.

Římov Reservoir.

The above-named problems belong to the category of questions of prediction or long term planning. Another category is the operative, dynamic water quality management. Whereas descriptive models (Biswas, 1981) are used for the first, prescriptive management models have to be developed for the latter. These are optimization models, including automatic decision criteria for selecting between alternative solutions. A dynamic optimization model based on AQUAMOD 1 and AQUAMOD 2 was published by Kalčeva et al. (1982) and Schindler and Straškraba (1982). As shown in Straškraba (1985), the model GIRL OLGA (Generalized Immitation of Reservoirs and Lakes–Optimum Low Growth of Algae) allows the investigation of new ways of eutrophication management.

Several management options were built into the model, the functioning of which was recognized on the basis of previous simulations. In addition to classical management strategies such as the reduction of inflow phosphorus, other biologically based, cheap interferences into the ecosystem were evaluated from the systems dynamics and economic point of view: epilimnetic mixing for decreasing light availability for algae, increase of light extinction or shading for the same reason, increase of phytoplankton control by zooplankton grazing through changes of fish populations, and utilization of multilayer outlets for decreasing phosphorus load of surface strata or increasing washing out of algae. The control philosophy is as follows:

Minimize price of management by finding a combination of alternatives which decreases eutrophication (measured as algal crop) just below some prescribed limit which is harmless for drinking water treatment. This is achieved by dynamic optimization, the algorithms being solved in cooperation with colleagues from the Institute of Theory of Information and Automatization, CAS Prague. Only some management options or portions thereof are of a dynamic nature; the others have an investment character. For both, restrictions of a biological and physical nature have to be respected during computations (dynamic optimization with state and parameter constraints). The model was applied to the situation in Želivka Reservoir for drinking water supply of Prague and it was recognized that it is possible to reach the desired drop of algal crop by the combination of only few options. Similar results were found for Slapy and Orlík reservoirs.

Based on models of this type, a new direction of applying ecological knowledge to practical problem solving was developed and coined "ecotechnology" (Straškraba, 1986). The approach is directed towards recognizing and testing new control possibilities. Mathematical models play an important role during the process. Another feature of ecotechnology is relying on ecosystem theory, particularly as to long-term effects of technology on nature (Straškraba, 1985).

Toward the Development of Aquatic Ecosystem Theory

Aquatic ecosystems are multicomponent systems with complicated webs of relations. The number of components, particularly of inoganic, dissolved and particulated organic chemical species as well of species of organisms can be as high as $10^4...10^5$. Sometimes only $10...10^2$ components and $10^2...10^3$ processes dominate as to their intensity of occurrence, but some compounds like vitamins are very active at very low concentrations. Other components and processes can be activated when conditions change—ecosystems are adaptive and self-organizing. Which one that would be, which species will occur and which material and energetical couplings will be realized depends largely on previous ecosystem history as well as on present internal state and external variables. Irreversible historical development is a characteristic ecosystem feature—an ecosystem is an evolving system. Fluctuations due to internal as well as external causes have a leading role in ecosystem evolution. Two sources of ecosystem state changes can be distinguished: deterministic and stochastic. Some portions of important processes can be described by chemical and physiological functions, the rest representing some "white noise." We have to understand, however, that what we consider random fluctuations at present can be either a superposition of several deterministic but presently unidentified processes, or real stochastic variability. It seems to be misleading to concentrate our effort of mathematical modelling on simply quantifying and elaborating ecosystems stochasticity without advancing towards understanding of the underlying processes and causes.

To summarize, the ecosystem is not a fixed structure fully deterministic "hard" system as we usually represent it in our models, but a soft, fuzzy system. Moreover, our possibilities for measuring, studying and quantitatively understanding the different ecosystem structures and their reactions on external and internal influences are rather limited, and it is not to be expected that such difficulties will be overcome in a few decades. The observation errors for ecological processes are higher than for purely physical ones; in addition, the natural processes rapidly fluctuate in time and space. We are unable to measure simultaneously more than a small fraction of the mutually interrelated processes and external disturbances.

One of the greatest difficulties of ecosystem studies is coupled with what will be called, in systems science, the multiresponse character. Multiresponse means that both the shape and parameters of a response to some factor is not the same under all circumstances. The reaction on one variable depends on additional influences. As an example, we may give the response of pelagic aquatic ecosystem to mixing. There may be no reaction of algal growth for an optically deep water body, but a very high one for an optically shallow lake. What is realized during the observation period is only a limited part of possible ecosystem behavior. If external conditions change as a result of human effects or if internal structure varies due, e.g., to immigration (or introduction) of new species, the ecosystem will behave rather differently. Usually we observe and measure ecosystems behavior during a restricted range of circumstances, such as during one or a few successive years. Reasonable agreement of such data and a model represented by fixed structure nonlinear ordinary differential equations does not per se guarantee a realistic estimate of future ecosystem changes. For these reasons, basic laws of physics and cybernetics have to be represented more thoroughly during development of aquatic ecosystems theory.

Thermodynamic Theory of Water Quality

Nonlinear thermodynamics of irreversible processes is applicable to all types of aquatic resources. From the thermodynamic point of view, these are open systems in energetic and material exchange with their surroundings. The energy and matter flow through the system cause the energy input and the output of entropy produced within the ecosystem. Biological processes, but also many chemical transformations, are nonlinear. The ecosystem is far from thermostatic equilibrium (which will represent a homogenized, dead state). Therefore, organized states, so-called "dissipative structures," are possible. Steady degradation of organization by respiration and death in the ecosystem has to be continuously balanced by new synthesis of organic matter for which supply of energy is necessary.

This led Mauersberger (1982a) to derive the shape of nonlinear relations of biological processes in the ecosystem from driving forces on the basis of thermodynamics of irreversible processes. An optimality principle is used which characterizes the deviation of the state of the biocoenosis from a stable stationary state. Because biological systems are able to "enlarge" or "extend" stable stationary states, he defined for the optimality principle a generalized excess entropy production, E. The time integral over a finite past time interval t should be minimized:

$$\int_{t-\tau}^{t} E \, dt' \Rightarrow \min ! \tag{9-3}$$

On this basis Mauersberger (1982a, 1985a) theoretically derived the process rates Y_{kn} as a function of driving forces X_{kn} (affinities):

$$Y_{kn} = K_{kn} [1 - D_{kn} e^{(-c_k X_{kn})}] \tag{9-4}$$

Here, k is the number of the species (k = 1,2,3, ...) and n the number of the process (n = 1,2,3,... for uptake, photosynthesis, assimilation, respiration, grazing etc.). The coefficients K_{kn} and D_{kn} reflect properties of the species, but also depend upon the stable stationary reference state used for the determination of the excess entropy production (Mauersberger, 1985a). Since from the entropy balance equation it can be inferred how the affinities X_{kn} are connected with state variables such as temperature, light intensity, nutrient concentrations, biomass etc. (Mauersberger, 1978, 1983), at least the type of the functional relationships between X_{kn} and these variables can be determined (Mauersberger, 1982b, 1984).

Cybernetic Basis of Ecological Modelling

In addition to external variables—light, temperature, nutrient supply—we have to respect internal control mechanisms, which lead, generally speaking, to survival and satisfactory development of species or populations. Of all species with a similar function in the ecosystem, those will be favored in each instant of time, that have properties best suited to the present systems state and external effects. The "suitability" is defined by means of optimality principles. Goal functions are defined differently by different authors. For individual organisms or single species populations, the term evolutionary "fitness" expresses the adaptive capability for survival (e.g. Wilson, 1980). More complex is the situation for interacting populations, e.g. prey-predators. Here the term coevolution started to be used to express the observed mutual adaptations of the interacting populations. For an ecosystem, more operative measures assumed to represent the consequences of the previous are used: optimum growth rates, maximum biomass. The adaptation to systems state by means of parameter changes is cybernetically described as self-adaptation, whereas structural change like substitution of dominating species is considered as self-organization (Straškraba, 1979). Until now it has been easier to imagine and formulate goals for optimization strategy of subsystems (species, populations) than for the whole system with complicated prey-predator interactions. A possibility for overcoming this difficulty is using hierarchical optimization or poly-optimization (Straškraba, 1980b). From the cybernetic point of view, the hierarchical structures in ecosystems can be distinguished not only in respect to biological organization (ecosystem, association, population, individuum, physiological process, cell) and to trophic levels, but also to control. Control can be imagined to switch from direct and feedback control to self-adaptation, self-organization and self-evolution as the disturbances become progressively more profound and time scales longer (Straškraba, 1983). Attempts of an exact formulation and suggestions for possibilities of numerical solutions are given by Bakule and Straškraba (1982).

A Cybernetic Model of Phytoplankton Structure

The first attempt to model self-adaptation of an aquatic ecosystem (Radtke and Straškraba, 1980; see also Straškraba and Gnauck, 1985) relies on the assumption that production and destruction processes of different phytoplankton species are dominantly determined by individual cell or colony size. As goal function the attainment of maximum possible phytoplankton biomass, B, over the period of one year is specified:

$$\int_{t=t_0}^{t} B \, dt \;\Rightarrow\; \max ! \tag{9-5}$$

For realizing equation 9-5, where B is determined by a dynamic phyto-plankton model with size dependent rates, a dynamic optimization procedure is used. It determines for each optimization time interval the optimum algal cell or colony size (or perhaps an average cell size of the phytoplankton population). This is viewed as the selection of species best fitting to present conditions. The relations of rate parameters to cell size were mostly approximated from empiri-cal observations.

Results of this experiment were encouraging in that automatic selection of algal "species" was obtained by the model. Also, changes of algal size were somewhat comparable with those observed in nature. It became evident that growth rates of the respective algae are not the most important parameters. The actual changes of biomass are determined as much by sedimentation and graz-ing as they are by growth rates. Also due to feed-backs in the system, e.g. through nutrient exhaustion and/or zooplankton growth, the advantages of a rapidly growing algal species are leveled out.

There were, however, many shortcomings in the underlying assumptions and simplifications of both the model and the solution algorithm. One particu-lar problem is the summing up of all algae into one compartment, the problem called by O'Neill (1979) transmutations across hierarchical levels. As shown by Bakule and Straškraba (1984), the aggregation of species based on Bell-man's Principle of Optimality is possible only when species overlap is neglect-ed, a simplification which is not always justified for dominating phytoplank-ters in nature.

Other problems are associated with the use of resulting biomass as the measure of fittingness of the phytoplankton associations. Although maximum biomass is one of the goal formulations (see next paragraphs), we have no direct mathematical nor biological proof that it is a feasible one.

Combination of Thermodynamic and Cybernetic Principles

We have shown that both the thermodynamic theory and the cybernetic approach use optimality principles (cf. Equations 9-3 and 9-5). It is therefore useful to compare the approaches in detail. It appeared that the goal functions used in both disciplines are different. It is to be noted that in mathematical biology rather different extremal principles and goals were applied, e.g. maximum energy flow (Lotka,1925; Odum, 1956), maximum entropy production (Glansdorff and Prigogine, 1971), maximum exergy (Jørgensen and Mejer, 1977), maximum organic substance (Whittaker and Woodwell, 1972), maximum biomass (Margalef, 1968). We have used in the thermodynamic direction a principle different from that one by Glansdorff and Prigogine: minimum generalized excess entropy production, and in the cybernetic direction the one by Margalef. In spite of this, the two approaches can be mutually complementary. The thermodynamic optimality principle can be formulated introducing continuously varying control variables (Mauersberger, 1985b). From this principle follow, inter alia, the functional relationships:

$$P = P_{max} (1 - e^{(-cZ_p)}) \geq 0 \qquad (9\text{-}6)$$

$$U = U_{max} (1 - e^{(-cZ_u)}) \geq 0 \qquad (9\text{-}7)$$

$$R = R^* (e^{(-cZ_R)} - 1) \geq 0 \qquad (9\text{-}8)$$

between the rates P, U, R of photosynthesis, uptake, respiration and the driving forces (affinities) Z_p, Z_u, Z_R. The dependence of these forces upon temperature, light intensity and nutrient concentrations can be inferred from the entropy balance equation. Determining the forces as functions of the controlling variables, the relationships between the rates P, U, R etc. and these variables are found. Mauersberger and Straškraba (1987) have shown that in this way it is possible to derive theoretically the relations of rates of physiological-ecological processes to phytoplankton cell or colony volume. At present such relations were used in cybernetic modelling based on empirical observations. The preliminary results obtained suggest the feasibility of this comparison but until now cannot be considered more than a methodical development.

Conclusions

The above review of the modelling activities in Czechoslovakia and GDR concerning shallow lakes and reservoirs demonstrates the need for combination of theoretical and applied model developments.

Present models were successfully applied to several actual water quality problems. Based on scenarios simulating water quality, management strategies were made for Lake Grosser Müggelsee and reservoirs in Czechoslovakia, GDR and USSR. By including costs for different management options, simulation models were extended to optimization models. They calculate explicitly the least cost strategies while conserving some water quality criteria.

Multiparametric dynamic optimization methods were used for numerical model solutions.

However, inherent inadequacies of the present modelling methodology were recognized. Ecosystems possess higher order dynamics (self-adaptation and self-organization capabilities) usually not covered by the models. Most functional relations are derived empirically, with less underlying theory. Two theoretical developments to overcome the difficulties are outlined: thermodynamic and cybernetic. Both are based on assumed optimality criteria for ecosystem behavior. One utilizes ideas derived primarily from physics and chemistry and the other one from economy and general systems theory. A recent attempt to compare the two specific approaches in detail proved that they can be mutually complementary. However, both independent development of the two approaches as well as their comparisons must continue.

Acknowledgments

The authors wish to thank the Institute of Landscape Ecology of the Czechoslovak Academy of Sciences, Cĕské Budĕjovice, Section of Hydrobiology, where most of this work was done. Thanks also to Dr. M. Dvořáková and Engr. L. Lhotka for collaboration during the development of mathematical models included in this chapter.

References

Bakule, L. and M. Straškraba. 1982. On multi-objective optimization in aquatic ecosystems. *Ecol. Modelling* 17:75-82.

Bakule, L. and M. Straškraba. 1984. Optimality in multispecies ecosystems. *Ecol. Modelling* 26:33-39.

Behrendt, H. and G. Schellenberger. 1987. Ein einfaches mathematisches Modell zur Simulation der Phytoplankton-und Sestonentwicklung in flachen Seen (EMSY). *Acta Hydrophysica*, (in press).

Biswas, A.K., ed. 1981. *Models for Water Quality Management.* McGraw Hill, New York.

Chytil, I. 1980. Simulation of river reservoirs and its application for radioactive pollutants. *Simulation of Systems in Biology and Medicine 1980.* Dum techniky CSVTS, Prague, Czechoslovakia.

Dvořáková, M. 1976. *Analytical model of the ecosystem of a standing water body.* Thesis, Charles Univ. Prague (In Czech).

Glansdorff, D. and I. Prigogine. 1971. *Thermodynamic Theory of Structure, Stability and Fluctuations.* John Wiley, New York.

Henderson-Sellers, B. and P.B.R. Archer. 1982. Models of annual cycles in lentic water bodies. *Hydrobiologia* 88:89-91.

Imboden, D.M., K. Lemmin, T. Joller, and M. Schuster. 1983. Mixing processes in lakes: mechanisms and ecological relevance. *Schweiz. Z. Hydrol.* 45:11-44.

Jørgensen, S.E. and H. Mejer. 1977. Ecological buffer capacity. *Ecol. Modelling* 3:39-62.

Kalčeva, R., J.V. Outrata, Z. Schindler, and M. Straškraba. 1982. An optimization model for the economic control of reservoir eutrophication. *Ecol. Modelling* 17:121-128.

Kozerski, H.P. 1977. Ein einfaches mathematisches Modell fur den Phosphoraustausch zwischen Sediment und Freiwasser. *Acta Hydrochim. Hydro-Biol.* 5:53-65.

Kozerski, H.P. and M. Dvořáková. 1982. A three layer aquatic ecosystem model: sediment submodel tests and some simulations. *Ecol. Modelling* 17:147-156.

Lotka, A.J. 1925. *Elements of Physical Biology.* Williams and Wilkins, Baltimore.

Margalef, R. 1968. *Perspectives in Ecological Theory.* Univ. Chicago Press, Chicago.

Markofsky, M. and D.R.F. Harleman. 1971. *A predictive model for thermal stratification and water quality in reservoirs.* MIT, Department of Civil Engineering, Ralph M. Parsons Laboratory for Water Resources and Hydrodynamics, Rept. No. 134, Cambridge.

Mauersberger, P. 1978. *On the theoretical basis of modelling the quality of surface and subsurface waters.* Proc. Baden Sympos., Sept. 1978, IAHS-Publ. 125:14-23.

Mauersberger, P. 1982a. Zur Bestimmung der nichtlinearen Beziehungen zwischen Raten und Affinitaten bei Produktions-und Abbauprozessen im aquatischen Okosystem. *Acta Hydrophys.* 27:125-130.

Mauersberger, P. 1982b. Rates of primary production, respiration and grazing in accordance with the balances of energy and entropy. *Ecol. Modelling* 17:1-10.

Mauersberger, P. 1983. General principles in deterministic water quality modelling. Pages 42-115 in G.T. Orlob, editor. *Mathematical Modelling of Water Quality: Streams, Lakes, and Reservoirs.* John Wiley, New York.

Mauersberger, P. 1984. Thermodynamic theory of the control of processes in aquatic ecosystems by temperature and light intensity. *Gerlands. Beitr. Geophys.* 93:314-322.

Mauersberger, P. 1985a. Optimal control of biological processes in aquatic ecosystems. *Gerlands. Beitr. Geophys.* 94:141-147.

Mauersberger, P. 1985b. Dominant controlling variables in the theory of biological processes in aquatic ecosystems. *Gerlands. Beitr. Geophys.* 94:161-165.

Mauersberger, P. and M. Straškraba. 1987. Two approaches to ecosystem modelling: thermodynamic and cybernetic. *Ecol. Modelling* (in press).

Nihoul, C.J., ed. 1975. *Modelling of Marine Systems.* Elsevier Oceanography Series 10, Elsevier, Amsterdam.

Octavio, M.K.A., G.H. Jirka, D.R.F. Harleman. 1977. *Vertical heat transport mechanisms in lakes and reservoirs.* MIT, Department of Civil Engineering, Ralph M. Parsons Laboratory for Water Resources and Hydrodynamics, Rep. No. 227., Cambridge.

Odum, H.T. 1956. Efficiencies, size of organisms and community structure. *Ecology* 37:592-597.

O'Neill, R.V. 1979. Transmutation across hierarchical levels. Pages 59-78 in G.S. Innis and R.V O'Neill, editors. *Systems Analysis of Ecosystems.* International Cooperative Publishing House, Fairland, Maryland.

Procházková, L. 1983. Dlouhodobé změny chemického složení vody Slapské nádrže. (Longterm changes of chemical composition of water in Slapy Reservoir; In Czech with English Summary). *Vodní Hospodářství, B,* 33:71-76.

Radtke, E. and M. Straškraba. 1980. Self-optimization in a phytoplankton model. *Ecol. Modelling* 9:247-268.

Schellenberger, G., H.P. Kozerski, H. Behrendt, and S. Hoeg. 1978. *A mathematical ecosystem model applicable to shallow water bodies.* Proc. Baden Symp., Sept. 1978. IAHS-Publ. 125:128136.

Schellenberger, G., H. Behrendt, H.P. Kozerski, and V. Mohaupt. 1983. Ein mathematisches Ökosystemmodell für eutrophe Flachgewasser. *Acta Hydrophys.* 28:109-172.

Schindler, Z. and M. Straškraba. 1982. Optimal control of reservoir eutrophication. (In Czech with English summary). *Vodohosp. časopis* SAV 30:536-548.

Straškraba, M. 1979. Natural control mechanisms in models of aquatic ecosystems. *Ecol. Modelling* 6:305-321.

Straškraba, M. 1980a. The effect of physical variables on freshwater production: analyses based on models. Pages 13-84 in E.D. LeCren, R.H. Lowe and McConnel, editors. *The Functioning of Freshwater Ecosystems.* Cambridge University Press, Cambridge.

Straškraba, M. 1980b. Cybernetic categories of ecosystem dynamics. *ISEM Journal* 2:81-96.

Straškraba, M. 1982. The application of predictive mathematical models of reservoir ecology and water quality. *Canad. Water Resource J.* 7:283-318.

Straškraba, M. 1983. Cybernetic formulation of control in ecosystems. *Ecol. Modelling* 18:85-98.

Straškraba, M. 1985. Managing of eutrophication by means of ecotechnology and mathematical modelling. Pages 17-27 in *Lakes Pollution and Recovery.* Proc. Internat. Congress, Rome, 15-18th April 1985.

Straškraba, M. 1986. Ecotechnological measures against eutrophication. Special Issue Limnology of Czechoslovak Reservoirs. *Limnologica* (Berlin) 17:237-249.

Straškraba, M., B. Desortová, and J. Fott. 1979. Zur Methodik der Bestimmung und Bewertung des Chlorphyll-a in Oberflächengewässer. *Acta Hydrochim. hydrobiol.* 7:569-590.

Straškraba, M. and A.H. Gnauck. 1985. *Freshwater Ecosystems: Modelling and Simulation.* Elsevier, Amsterdam, 309 pages.

Straškraba, M. and V. Straškrabová. 1975. Management problems of Slapy Reservoir, Bohemia, Czechoslovakia. Proc. of the Symp. on effects of storage on water quality. *Reading 1975*: 449-484.

Straškrabová, V. 1975. Self-purification capacity of impoundments. *Water Res.* 9:1171-1177.

Uhlmann, O. 1972. Das Staugewässer als offenes System und als Reaktor. *Verh. Internat. Verein Limnol.* 18:761-778.

Whittaker, R.H. and G.M. Wodwell. 1972. Evolution of natural communities. Pages 137-156 in J.A. Wiens, editor. *Ecosystem Structure and Function.* Oregon State Univ. Press, Corvallis, Oregon.

Wilson, D.S. 1980. *The Natural Selection of Populations and Communities.* The Benjamin/Cummings Publ. Co., Menlo Park, California.

10/ MODELLING EUTROPHICATION OF SHALLOW LAKES

Sven E. Jørgensen

Many shallow lakes suffer from eutrophication due to high inputs of nutrients relative to the water volume. It is therefore not astonishing that many eutrophication models have been developed for shallow lakes. This chapter will attempt to present the state of the art in modelling the eutrophication process for shallow lakes today by offering a summary of what we learned from our experience in this field during the last 12 years. Also presented are some of the latest results developed by the author of this chapter in modelling eutrophication of shallow waters.

What Have We Learned?

Modelling the eutrophication of lakes was initiated in the late 1960s by researchers such as Orlob and Chen (see, e.g., Chen and Orlob, 1972, 1975). Many of the processes that were described in the first eutrophication models were not sufficiently known at that time, but modelling is an iterative process. The system, however, is too complex to enable us to give a complete and accurate description in all detail. It is always possible to improve our modelling description, but we shall never reach a final complete version of the model which cannot be improved in some way at a later stage. The first eutrophication models can easily be criticized today, but it should be recognized that it was not possible to employ knowledge which was not yet developed about processes relevant to eutrophication. To their credit, the first models served as very useful management tools and as indicators for areas of concentration in future research to improve models of eutrophication.

It was clear from the experience gained during the early 1970s that we needed better descriptions of several processes of importance for the eutrophication of lakes. For example, the exchange of nutrients between sediments and water is a significant process in shallow lakes, as a very substantial amount of the total nutrient content is stored in the sediments. The internal loading may be greater than the external loading of nutrients, such as in cases where the external loading is reduced significantly and the sediments still contain high nutrient concentrations. Today we do have a better knowledge of this process and can give better descriptions in the eutrophication models developed and applied most recently (see, for example, the review of this process by Kamp-Nielsen, l983; Lijklema, 1980, 1983).

The description of phytoplankton growth has also been widely discussed in the field of modelling during the last decade and two general approaches have been applied to phytoplankton-nutrient interactions. The first approach involved a Monod-type equation which relates the rate of growth with the nutrient concentration. The second approach is a two-step model which involves the uptake of nutrients related to the nutrient concentration in the water and the growth controlled by the intracellular concentration of nutrients. Our experience of using the two different approaches has clearly shown that the more complex two-step description is needed in shallow very eutrophic lakes, while it is rarely necessary to use this complex approach in other cases.

Much research has been devoted to gathering knowledge on the important process of grazing, but there is less understanding of the usefulness of various process equations in modelling eutrophication. It may be revealed in the future (where we shall be very concerned about the right methods to describe the changes in ecological structure observed in lakes and other ecosystems) which grazing expressions are most useful to apply to eutrophication models. Similarly, our experiences are limited on the role of fish in the eutrophication process and the inclusion of predation of fish in these models.

At the beginning of the 1970s, many modellers were so overwhelmed by the new possibilities that computers had given to us that they made the models very complex, more complex than the knowledge and data could support. Today we have learned how to find a balance in model complexity (which considers the data availability), our ecological knowledge about the processes involved, the scope of the model, and the modelled ecosystem. This has probably been one of the most pronounced advances in ecological modelling: how to assess the right complexity of a model considering all the above mentioned factors. In this context we should mention that the dynamics of the processes included in a model are crucial for the data program set up. If there is accordance between the two, it is possible to make predictions on the dynamics of the processes, but if such an accordance is not present, it should not be expected that the model can give such predictions. The dynamics of phytoplankton growth are such that almost daily measurements are needed at least during the bloom periods. If such dense measurements are not available, it must not be expected that a model that has based its calibration on these data can give accurate predictions on the maximum phytoplankton concentration and the time at which it occurs. This problem is treated in details in Jørgensen and Mejer (1979) and in Jørgensen et al. (1981).

Through the use of models — in science or management — we have come to know lakes as systems much better. This again has made it easier to understand which processes are of importance in lake management models; because of this experience we are much more able to construct the right models. We have, furthermore, gained confidence in the use of models for environmental management because we have been able in some case studies to observe and evaluate the applicability of models. The case study which is presented below is one of the few case studies where the prognosis has been validated (It is probably one of the only models where the prediction was published before the data for the validation actually was collected.)

In the late 1960s, when modelling of the eutrophication process began, many biologists and ecologists questioned whether it was possible to model such a complex system as a lake. They claimed that so many processes were directly or indirectly involved in eutrophication that it was impossible to construct a model of even a limited reliability. It is possible, looking at the experience gained during the last 15 years, to conclude today that it is feasible to model eutrophication of a lake or reservoir and to use it for a better scientific understanding of the ecosystem or for management of lakes and reservoirs. But our experience has also shown us that the formulation of processes into mathematics and the translation of mathematical expressions into a suitable computer language are not the difficult parts of modelling. It is a far more complicated task to get the ecological data and the ecological knowledge that are needed to be able to make the mathematical formulations.

We have learned through modelling that ecological systems have many types of regulation and feedback mechanisms which make it very difficult to consider all the changes that we may observe in ecosystems. Therefore, we have still not been able to construct models which are able to predict changes in species composition or in ecological structure. Some first approaches have been taken in this regard; it will most probably be possible to construct such models within a few years. This will improve the eutrophication models considerably, as the composition of phytoplankton species changes when the nutrient loading is changed, a phenomenon not accounted for in present models.

It must be admitted that we have gained comprehensive experience in modelling the eutrophication process during the last 15 years, but we are in the center of a very fast development. So it must be expected that many new results will emerge during the coming years in the field of modelling the eutrophication process.

A Eutrophication Model for a Shallow Lake

The model that will be presented in this section has been under development since 1974. The individual steps in this development are illustrated in Figure 10-1. It illustrates the general development in the field, the difficulty of getting all the needed data, and the ecological knowledge required to construct a good

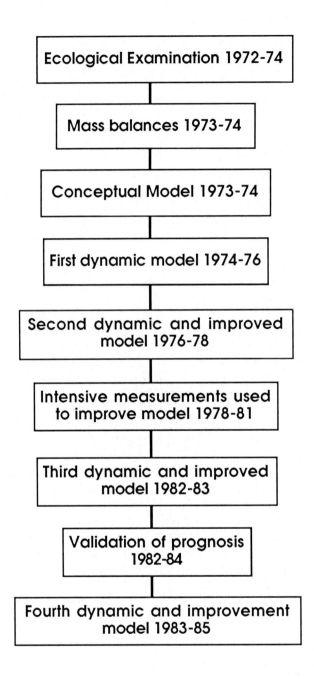

Figure 10-1. General development schedule for eutrophication model discussed in text.

workable model. As seen from the figure, the development in 1972-76 followed the usual recommended pattern. A pre-examination was required to get an ecological overview of the ecosystem and the processes of importance for the problem in focus. Mass balances give a good idea of the mass flows and where the significant processes take place in the ecosystem. It is always necessary to set up a conceptual model to structure the ideas behind the modelling activity. Finally , the result of this work, a dynamic model, was set up. The calibration was, however, not satisfactory and therefore it was necessary to test other ideas, including the two-step model of phytoplankton growth. This led to a second improved dynamic model which gave a much more satisfactory calibration and later also an acceptable validation. As discussed in the last section, it was necessary to base an exact prediction of the phytoplankton concentration at maximum on dense measurements. These were carried out in 1978 and in following years, making it possible to improve the calibration further and also leading to a new and better expression for temperature control of phytoplankton growth.

This model has not only been validated, but it has also been possible to validate a prediction which was published beforehand. This gave the opportunity to work further with the model and introduce a few improvements. The question of how to consider the possible changes in species composition is, however, still not solved, although some promising results have been obtained (see Jørgensen, 1986). This stepwise development of a eutrophication model can be found in detail in the literature by Jørgensen et al. (1973), Jørgensen (1976), Jørgensen et al. (1978), Jørgensen et al. (1981), and Jørgensen et al. (1986).

The characteristic features of the model may be summarized as follows:

1) The phytoplankton growth is described as a two-step process as described above.

2) The sediment-water exchange processes are described in more detail than in most eutrophication models. This submodel distinguishes between a part of the nutrients which is exchangeable and another part which is nonexchangeable. The exchange process is described as a two-step process: decomposition of the organic matter accompanied by release of nutrients to the interstitial water and diffusion from the interstitial water to the lake water. The submodel is illustrated in Figure 10-2.

3) The other state variables of the model can be seen from the conceptual diagram of the phosphorus part of the model (Figure 10-3). Similar processes and state variables are included for carbon and nitrogen. Notice that the three elements have independent cycling as a consequence of the two-step growth description for phytoplankton.

4) It is often possible to find several alternative expressions for the same process. In most cases the various alternatives have been tested in the calibration phase to assure that an acceptable equation is used in this context.

5) The model has been used in 16 different case studies with some modifications which can be explained by the ecological characteristics of each ecosystem. A comparison of the values obtained for the parameters during cali-

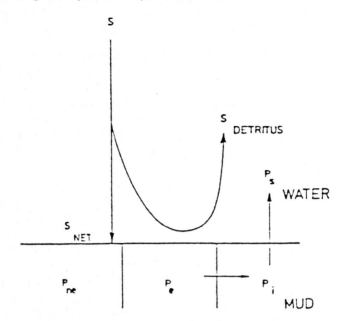

Figure 10-2. Sediment-water phosphorus exchange submodel. S = sedimentation, which divides into S_{NET}, sediment actually carried to the sediment, and $S_{DETRITUS}$, sediment mineralized by microbial activity in the water column; P_{ne} = non-exchangeable phosphorus in unstabilized sediment; P_e = exchangeable phosphorus in unstabilized sediment; P_i = phosphorus in interstitial water; P_s = dissolved phosphorus in water.

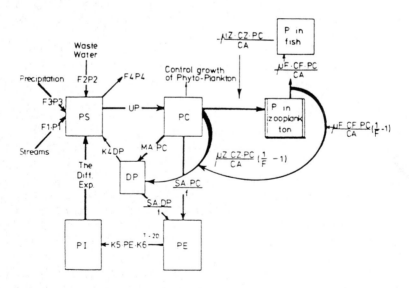

Figure 10-3. Phosphorus cycle in eutrophication model discussed in text. PS = soluble phosphorus; PC = algal phosphorus; DP = detritus phosphorus; PI = interstitial phosphorus; PE = exchangeable phosphorus in sediment.

bration for the different case studies shows that the variation from case to case is rather limited and at least within the range of values which can be found in the literature.

A Case Study: Lake Glumsø

The investigations of Lake Glumsø in Denmark have been the most detailed among the above mentioned 16 case studies. This lake was chosen for a very detailed test of the possibilities for development of a eutrophication model because it has the following advantages:

1) The lake is small (area 0.266 km² and average depth 1.8m), giving it a simple hydrology. This means that it was possible to focus on the ecological processes and not go into details on the hydrology.

2) When the investigations started in 1972, a reduction of nutrient loading was planned, giving the possibility of validating the model prediction—a very important part of the entire project. This reduction did, in fact, take place. Until April 1981, wastewater from the village Glumsø was discharged to the lake after a mechanical-biological treatment. After April 1981, the wastewater was discharged downstream of the lake. This implied a reduction in the phosphorus loading of 73%.

3) The lake was hypereutrophic, implying that a reduction in the loading most probably would give a clear result.

4) The retention time for the water was short (3-6 months), yielding results of the reduction in nutrient loading within short time.

Table 10-1 shows mass balances, nutrients concentrations, and phytoplank-

Table 10-1. Mass balances and nutrient and phytoplankton concentrations for Lake Glumsø, Denmark, 1972-1985.

	1972	1975	1981	1982	1983	1984	1985
Phosphorus load, g P/m²-yr	4.5	6.0	1.6	1.6	(1.6)	(1.6)	(1.6)
Nitrogen load, gN/m²-yr	36.1	64	41	--	--	--	--
Average P conc., g P/m³	1.27	1.44	1.12	0.75	0.51	0.57	0.43
Days PO_4^{-3} conc.<30 mg/m³	0	0	0	27	40	71	--
Days inorg. N < 100 mg/m³	0	130	68	170	152	--	--
Average chlorophyll, mg/m³	812	549	641	301	344	290	192
P retention, g P/m²-yr	0.5	1.2	-4.0	-1.0	(+)	(+)	(+)
Denitrification, g N/m²-yr	15	37	--	--	--	--	--

ton concentrations for selected years during the period 1972-1985. Notice that there is negative net sedimentation of phosphorus in 1981 and 1982, which means that the release of phosphorus from the sediment has been greater than the phosphorus removed from the lake water by sedimentation. Figure 10-4 shows how the chlorophyll concentration has changed with the reduced phosphorus concentration from 1981-1985. The figure compares this relationship for nordic lakes and all OECD-lakes. Lake Glumsø follows the Nordic lakes and is above the chlorophyll level for the OECD-lakes.

Table 10-2 contains the prognosis which was already published in 1976. Table 10-3 gives a comparison of measured and predicted values. The accordance is believed to be satisfactory. Figures 10-5 and 10-6 show the seasonal changes in nitrogen, phosphorus, and phytoplankton concentrations; predicted as well as measured values are shown.

Conclusions

It is possible, based on the results of this case study, to conclude that it is feasible to develop and construct eutrophication models and use them as management tools to make predictions. This does not imply that the model cannot be improved considerably. The model which has been used assumes a certain ecological structure and certain species of phytoplankton and zooplankton. It was observed, however, that a considerable shift in the phytoplankton species composition took place. Furthermore, it is also desirable to improve the prognosis. A correlation coefficient for a comparison of observed and predicted values was 0.66; it would probably be desirable to get a correlation coefficient of 0.85 or more.

It seems possible to make such improvements of the model and to include the

Table 10-2. Lake Glumsø model predictions for primary productivity and minimum transparency for two different phosphorus concentrations in wastewater: Case A: 0.4 mg P/l; Case B: 0.1 mg P/l.

	Third Year		Ninth Year	
	Case A	Case B	Case A	Case B
Productivity, g C /m^2-yr	650	500[a]	500	320[a]
Minimum transparency, cm	50	60	60	75

[a] an error of 3 percent on these values could be expected if the validation holds.

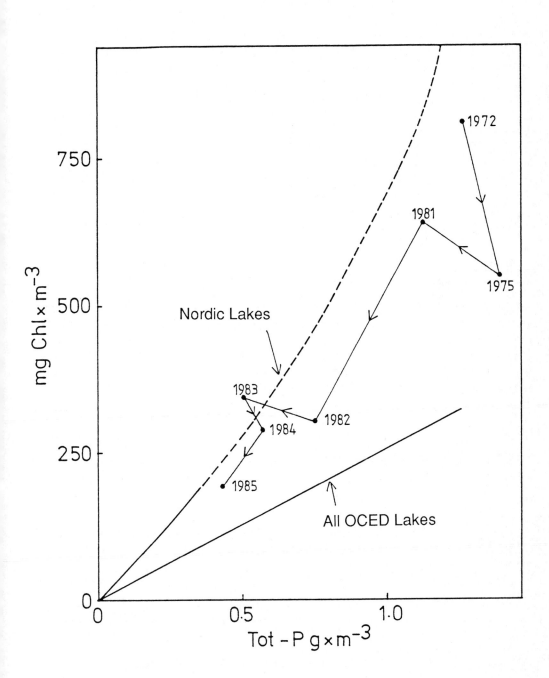

Figure 10-4. Chlorophyll concentration changes in Lake Glumsø, Denmark from 1972 to 1985 (solid line with arrows). Data are compared with relationships developed by OECD for nordic lakes (dotted line) and all OECD lakes (solid line).

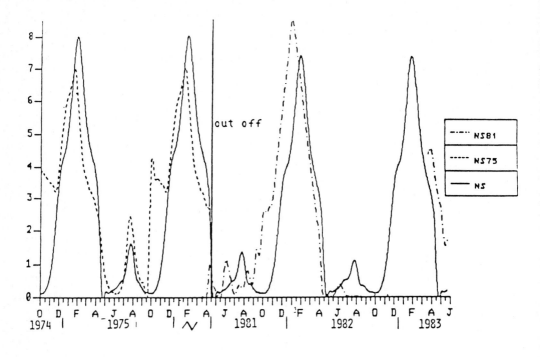

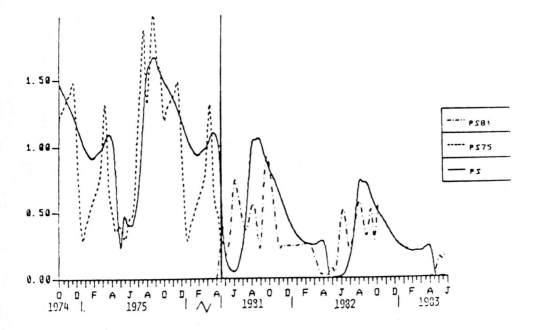

Figure 10-5. Model prognosis validation with 1974-75 used for validation, and 1981-83 data used for validation of the prognisis for a) nitrogen, and b) phosphorus. Solid line is model prediction, dotted lines are field data for 1975 and 1981.

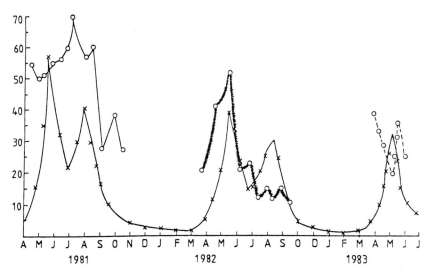

Figure 10-6. Model prognosis validation for phytoplankton chlorophyll concentration (mg/m³) with circles corresponding to measured values and crosses corresponding to model output.

Table 10-3. Comparison of model predictions (Case A: 92 percent reduction in phosphorus inflow) with actual measurements (actual case: 88 percent reduction in phosphorus inflow).

	Model Prediction	Measurement
Minimum Transparency, cm		
first year	20	20
second year	30	25
third year	45	50
Maximum Primary Productivity, g C/m²-day		
first year	9.5±0.8	5.5±0.5
second year		
spring	6.0±0.5	11±1.1
summer	4.5±0.4	3.5±0.4
autumn	2.0±0.2	1.5±0.2
third year	5.0±0.4	6.2±0.6
Chlorophyll, Spring Maximum, mg/m³		
first year	750±112	800±80
second year	520±78	550±55
third year	320±48	380±38

changes in species composition by use of a control variable to predict the changes in parameters as a consequence of the changes in the species composition. It is impossible to give all the details of these considerations. Those who are interested to learn more about it are referred to Jørgensen and Mejer (1979, 1981) and Jørgensen (1982, 1986).

References

Chen, C.W. and G.T. Orlob. 1972. *Ecological simulation for aquatic environments*. Office of Water Resources Research Report C-2044. WRE-1-0500. U.S. Department of Interior, Washington, D.C.

Chen, C.W. and G.T. Orlob. 1975. Ecological simulation for aquatic environments. Pages 476-587 in B. C. Patten, editor. *Systems Analysis and Simulation in Ecology, Volume III*. Academic Press, New York.

Jørgensen, S.E. 1976. A eutrophication model for a lake. *Ecol. Modelling* 2: 147-165.

Jørgensen, S.E. 1982. Exergy and buffering capacity in ecological systems. Pages 61-72 in W. J. Mitsch, R. K. Ragade, R. W. Bosserman, and J. A. Dillon, Jr., editors. *Energetics and Systems*, Ann Arbor Science, Ann Arbor, Michigan.

Jørgensen, S.E. 1986. *Fundamentals of Ecological Modelling*. Elsevier, Amsterdam. 389 pages.

Jørgensen, S.E., O.S. Jacobsen, and I. Hoi. 1973. A prognosis for a lake. *VATTEN* 29:382-404.

Jørgensen, S.E., H. Mejer, and M. Friis. 1978. Examination of a lake model. *Ecol. Modelling* 4:253-278.

Jørgensen, S.E. and H. Mejer. 1979. A holistic approach to ecological modelling. *Ecol. Modelling* 7: 169-189.

Jørgensen, S.E. and H. Mejer. 1981. Exergy as a key function in ecological models. Pages 587-590 in W. J. Mitsch, R. W. Bosserman, and J. M. Klopatek, editors. *Energy and Ecological Modelling*, Elsevier, Amsterdam.

Jørgensen, S.E., L.A. Jørgensen, L. Kamp-Nielsen, and H.F. Mejer. 1981. Parameter estimation in eutrophication modelling. *Ecol. Modelling* 13: 111-129.

Jørgensen, S.E., L. Kamp-Nielsen, T. Christensen, J. Windolf-Nielsen and B. Westergaard. 1986. Validation of a prognosis based upon a eutrophic model. *Ecol. Modelling* 32:165-182.

Kamp-Nielsen, L. 1983. Sediment-water exchange models. Pages 387-420 in S. E. Jørgensen, editor. *Application of Ecological Modelling in Environmental Management, Part A*. Elsevier, Amsterdam.

Lijklema, L. 1980. Interaction of orthophosphate with iron [III] and aluminum hydroxides. *Environmental Science and Technology* 14: 537-541.

Lijklema, L. 1983. Internal loading. *Water Supply* 1:35-42.

11/ SYSTEMS ECOLOGY OF OKEFENOKEE SWAMP

Bernard C. Patten

A vast inaccessible wetland wilderness, Okefenokee Swamp is being studied as a hierarchical, dynamical system. Formal system theory, which underlies empirical research and modelling at different levels of organization, makes possible lawful, in addition to empirical, development of findings. The study and modelling plans are outlined and illustrated. To emphasize the difference between systems and non-systems approaches, the significance of the concept of vegetation climax under both is examined. The climax as an ideal has meaning in systems ecology, whereas in absence of physical realization it has none in normal ecology.

Okefenokee Swamp is one of the last great vestiges of American wilderness left in the eastern United States. A magnificent wetland ecosystem of the Southeastern Coastal Plain, its primeval character is protected as a National Wildlife Refuge and Wilderness Area. The swamp, more than half the size of Rhode Island (1754 km^2 in a 3781 km^2 watershed), has been studied, using a systems approach, by the University of Georgia under U.S. National Science Foundation (NSF) auspices since 1975. This paper will give some perspectives on the approach, how it has helped to organize the study, what has come out of it, and what is in prospect.

The purpose of the study is to understand Okefenokee as an organized whole, and in the process to generate some new avenues of ecology from the systems approach. There are several elements to the approach: (1) an overall plan which has identified ultimate goals and the steps to reach them; (2) a basis in mathematical system theory for all theoretical and empirical work; (3) a modelling plan consistent with this basis; and (4) empirical research, initially loosely guided by the basis and models, but having a tightening relationship to them as the program matures. Only a cursory description of these elements is possible here, with a few examples; the theoretical rather than empirical side

189

will be emphasized as this is what is most different among ecosystem studies.

Figure 11-1 depicts the Okefenokee watershed, consisting of upland and palustrine portions. The uplands rise to 56 m in the northwest, and general elevation of the swamp is 37 m. Hydrologically, the system is ombrotrophic. The situation is complex, however, and future research may alter the present assessment. Eighty-three percent of precipitation falling on the uplands is lost as evapotranspiration, whereas only 54% of swamp rainfall leaves the watershed by this means. Twenty-two percent of upland precipitation flows to the swamp as surface streamflow. Two streams flow out of the watershed, the Suwannee and St. Mary's Rivers (Figure 11-1), which account respectively for 88% and 12% of water lost as streamflow, namely 11% of upland precipitation and 45% of swamp precipitation. Rainfall on the uplands has a future residence time of 1.45 years in the watershed, while swamp precipitation remains only 136 days. The composition of surface water in the swamp is 18% derived from upland rainfall and 82% from rain on the swamp. Swamp subsurface water, however, originates 99% in upland precipitation and only 1% as swamp precipitation. Swamp subsurface water will remain in the watershed for an average of 14 years.

The hydrologic vital statistics just given were not provided only because swamps live and die by the water they receive. These data should be recognized as not the kinds of numbers usually seen in ecosystem studies. In fact, they were derived from a water budget model (Patten and Matis, 1982) by a new systems analysis methodology, *environ analysis* (Matis and Patten, 1981), which implements an ecological system theory of environment (e.g., Patten et al., 1976; Patten, 1978, 1982; Patten and Auble, 1981). Environ theory and analysis are both expressions of the same general system theory that forms the Okefenokee program basis (see below). The water model referred to has been discussed elsewhere (particularly in Patten and Matis, 1982 and Patten, 1982), but later in this paper another similar model will be presented to illustrate environs.

Study Plan

Three stages of development comprise the program plan. The study would have to be multidisciplinary and dominantly empirical for a long time because little modern ecological knowledge of the Okefenokee existed at the outset. The first stage, now nearing completion, was to initiate field studies on a variety of topics to spawn a future, but at the same time establish a systems orientation in the program that would persist. This was accomplished through a set of Ph.D. dissertation studies by systems ecology students of the author (Rykiel, 1977; Bosserman, 1979; Blood, 1981; Auble, 1982; Hamilton, 1982; Flebbe, 1982; Barber, 1983; Stinner, 1983; and Glasser, 1986). Each study, a field investigation comprehensively drawn and data-rich, included some conceptual or mathematical modelling, or fit the modelling context described below. The collective result left an indelible systems mark on the program.

These first studies opened up many new areas for future research by

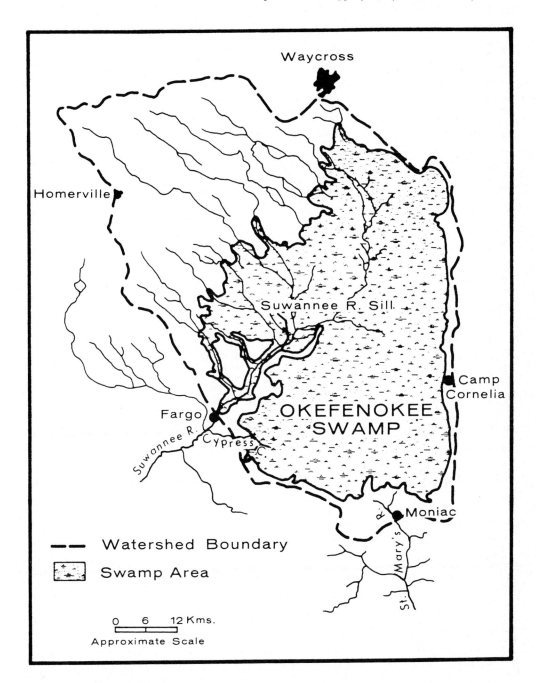

Figure 11-1. The Okefenokee swamp-upland watershed. Most of the upland (non-palustrine) area is in the northwest quadrant. Upland input streams and the two output streams, the Suwannee and St Mary's Rivers, are shown together with the Suwannee River Sill. Cypress Creek is a major output stream that flows into the Suwannee River (from Patten and Matis, 1982).

raising interesting questions and delineating areas of ignorance. Stage two was designed to expand the intellectual base of the program by including non-systems ecologists—faculty associates, postdoctorals and collaborators from other institutions. The systems work during this stage consists of modelling, loosely coordinated with the dominant empirical research, and development of ecological system theory for testing and expression in the third stage.

During stage three, empirical studies would still continue, but now more closely motivated by systems-theoretical questions and the need to qualify and quantify definitive models under development. "Integrated Studies of the Okefenokee Swamp Ecosystem" is the name as well as the goal of the program. An evolution toward tightening integration of more and more refined knowledge through the interplay of theory, modelling and empirical work is the plan.

System Theory Basis

Most mathematical system theories (e.g., Gill, 1962; Klir, 1969; Wymore, 1967; Windeknecht,1971; Mesarovic and Takahara, 1975; and Zadeh and Desoer, 1963) have greater or lesser similarities. The last two indicated are the deepest and most formal, and it is these along with hierarchy theory (e.g., Patten, 1973; Mesarovic, Macko and Takahara, 1970) that provide the basis for the swamp systems program.

Ecological systems have environments and are therefore open and can be treated as input-output systems in which a determinate (unique) relation is established between input and output through a "dynamical system" (Mesarovic and Takahara, 1975) or "state space" (Zadeh and Desoer, 1963) representation:

$$\rho: Z \mathrm{x} X \rightarrow Y, \quad \phi : Z \mathrm{x} X \rightarrow X. \tag{11-1}$$

Set function here corresponds to ordinary functional notation, namely $y = \rho(z,x)$, $x = \phi(z,x)$, where x, y and z are scalar or vector variables assigned values from the respective sets X, Y and Z. The response function ρ takes values for input variables from Z and state variables from X (the state space), and maps them into values Y of output variables. The state transition function provides for state changes through time.

As an example, vegetation in the Okefenokee Swamp is influenced by fire and water. Let fire be an input variable Z1, taking values (e.g., light, moderate, severe, etc.), from a set Z. Vegetation can be represented by an output variable y that assumes values (e.g., undamaged, slightly damaged, severely burned, etc.) from another value set Y. If water level is high, then fire damage will tend to be lighter than otherwise; water level mediates the vegetation's response to fire, and therefore becomes a state variable x to which values are assigned from X (say cm above mean sea level). The swamp's response to fire can then be expressed by a response function ρ (Equation 11-1) which maps fire severity z1, and water level x into vegetation response y. Furthermore, water level is a function of precipitation and current water level. Precipitation can be represented by a second input variable z2 which takes values (e.g., mm of rainfall, etc.), from the input value set Z. Therefore, water level in the swamp can be generated dynamically by a transition function φ (Equation 11-1) which takes precipitation z2 and water level x into subsequent water levels.

Some general rules follow. Given two identical systems (same lawful behavior as defined by identical r and f functions): (1) identical inputs and states will yield identical outputs, (2) identical inputs but different states will yield different outputs, and (3) different inputs with identical states will also yield different outputs. "Identical" and "different" can be taken within a stochastic context if desired. That is, dynamical systems can be statistically determined as well as deterministic, the latter simpler case being under discussion. Figure 11-2 (from Hamilton 1982, p. 45) illustrates generalized effects of fire (input) and water level (state) on the composition of Okefenokee vegetation (output). If fire does not occur (zl) and water level remains above the peat surface (x1) , then cypress (*Taxodium*)—bay (*Magnolia, Persea, Gordonia*)—blackgum (*Nyssa*) swamps (y1) are replaced by mixed hardwoods (y2) containing blackgum, bays, maple (*Acer*), ash (*Fraxinus*), oak (*Quercus*) and sweetgum (*Liquidambar*) (Figure 11-2A). Light or moderate fires (z2) under these water level conditions, i.e. a different input combined with the same state, are required to maintain the cypress-bay-blackgum stage (Figure 11-2B). When water levels are at or below the peat surface (x2), corresponding to a different state, light to moderate fires (z2) produce bay swamps (y3) (Figure 11-2C), whereas moderate to severe fires generate·cypress-shrub (*Lyonia, Cyrilla, Ilex, Itea, Leucothoe*) swamps (y4) (Figure 11-2D).

In describing Okefenokee plant succession in dynamical system terms, Hamilton (1982) in effect produced a system theory of succession. This is the third ecological topic formulated around the response and state transition function pair (the first two were environment, Patten, 1978, and niche, Patten and Auble, 1981), attesting to the generality of the state space model as a basis for systems ecology theory. In ecosystem appli·ations, the dynamical system is usefully considered to be a member of a larger dynamical system, and at the same time to be made up of smaller dynamical systems, that is, to be an object in a discrete nested hierarchy. The state space model is readily applied at any level of organization. Hierarchical dynamical systems, then, form the system theory basis of the Okefenokee study.

Modelling Plan

Zeigler (1978) stressed the need for diverse modelling approaches in studies of large, complex systems. A modelling program should unfold in different phases, encompass different levels of organization, implement multiple viewpoints provided by different experimental frames, and admit different kinds of mathematical formalisms. Figure 11-3 shows the initially formulated plan for the Okefenokee study.

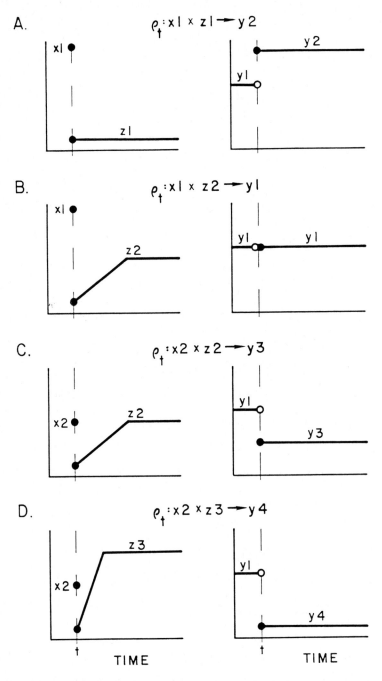

Figure 11-2. Species composition of Okefenokee vegetation (output, **y**) as functions of water level (state, **x**) and fire (input, **z**). Water level may be above (x1) or below (x2) the peat surface. Fire may be absent (z1), light to moderate (z2) or severe (z3). Vegetation types are cypress–bay–blackgum (y1), mixed hardwood (y2), bay swamps (y3), and cypress–shrub swamps (y4) (from Hamilton, 1982).

INTEGRATED STUDIES OF THE OKEFENOKEE SWAMP ECOSYSTEM
PLAN FOR MULTIPHASE, MULTILEVEL, MULTIFRAME, MULTIFORMALISM MODELING

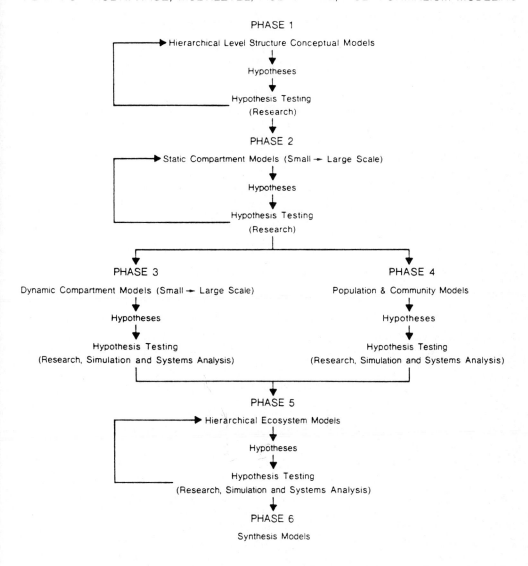

Figure 11-3. Phased modelling plan for the Okefenokee study.

In phase 1, a series of hierarchical conceptual models would be produced to guide research at different levels and provide a holistic context for assimilating results of this research. The current model serving this purpose will be outlined in the next section. Phase 2 was motivated by the assumption that data would be sparse, particularly early in the program, and only small models could be supported. The water model referred to earlier was one of these. Their purpose would be to summarize acquired information, extend this through model analyses, and in the process identify further research questions. Also, they would provide subjects on which to explore new analysis methodologies, such as environ analysis. One of the models in this group, comparing carbon biogeochemistry in three Okefenokee habitats, will be discussed below. Phase 3 visualizes extension of the static, second phase models to dynamic simulation models. None of these has yet been produced, although hydrology knowledge is now advancing to the point where dynamic considerations are needed (e.g., Blood, 1981), and can be supported by available data. Phase 4 consists of population and community models. Hamilton's (1982) representation of Okefenokee succession is one of these. Barber (1983) has completed another, a population model of orb weaving spiders in Okefenokee shrub swamps. A third currently under development is a stochastic stand model of woody vegetation dynamics, sensitive to water level and fire perturbations (Glasser and Barber, 1983). Phases 5 and 6 are for the future.

Hierarchical Organizing Model

The current version of the phase 1 hierarchical model is illustrated in Figure 11-4. This model serves as the focal point for specific research efforts. Hierarchy theory is built in. Specifically, the higher the level the larger the relevant space scales and the longer the relevant time scales. This information determines sampling specifications in terms of spatial distribution and frequency, and in many different ways helps to set research priorities and design studies.

The Geographic Region (level 0) provides a regional context in which information on climate, meteorology, geology, soils and hydrology can be developed. Particular emphasis has been given to the question of the swamp's geologic origin, and the pursuit of this problem has ranged far outside the immediate Okefenokee region. The latter is defined by the Watershed Ecosystem (level 1, Figure 11-1) because of the importance of hydrology. Surface divides bound this region. Subsurface divides may not correspond to these, however, because strata under the swamp rise to the surface in the northwestern uplands making recharge of subsurface aquifers outside the surface divide possible.

The watershed, defined as it is, is unusual in encompassing two drainages, the Suwannee and St. Mary's. These cannot be delimited presently within the swamp, and so another concept, Input Drainage (level 2), is used to define the next level. A drainage is a topographic region drained by a stream. An "input drainage" is the same only defined by stream inputs. The distinction is

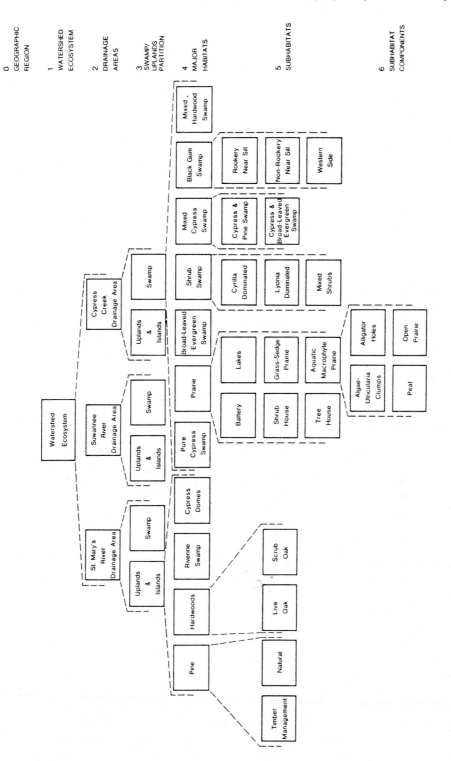

Figure 11-4. Hierarchical model for the Okefenokee regional ecosystem.

motivated by environ theory. A drainage corresponds to an input environ (which generates an output, namely an outflowing stream), whereas an input drainage (generated by an inflowing stream) corresponds to an output environ. The northwestern input drainage of the Okefenokee is fed by the entire collection of inflowing upland streams (Figure 11-1), whereas the southeastern input drainage is not. Deeper waters and faster surface flows characterize the northeastern subsystem, which is bounded from the southeastern roughly by the middle fork of the Suwannee River (the stream shown in the upper (north) central section of the swamp in Figure 11-1.

Each input drainage is subdivided into an upland and palustrine component to form a Swamp/Upland Partition (level 3). Cultivated slash pine (*Pinus elliottii*) dominates the uplands, with riverine swamps occurring in bands along stream channels. Mixed cypress forests (domes) occupy local depressions formed by sinkholes (doleens) in the karsted substrate. Major communities (level 4) consist of pine, hardwoods, riverine swamps and cypress domes on the uplands, and seven palustrine communities: marshes (locally called "prairies") of two types, aquatic bed (macrophyte) and emergent persistent (grass-sedge), which occupy 31% of the swamp; shrub swamps, among the densest vegetation in North America (up to 6.73 x 10^5 stems per hectare (J.E. Glasser, pers. comm.); only 6% of sunlight penetration to 45.7 cm above the ground surface, Hamilton,1982), covering 37%; broad-leaved evergreen forests dominated by the three fire sensitive bay species; blackgum forests, restricted to areas adjacent to the northwestern uplands lumbered for cypress in the first quarter of this century, and together with bay forests comprising 7% of the swamp; mixed cypress, with bays and blackgum as canopy species; pure cypress, limited to deep waters of the northwestern input drainage, and together with mixed cypress accounting for 25% of the swamp area; and mixed hardwoods, the regional climax (Monk, 1968) represented in the swamp only by individuals of some of the characteristic species. Relationships between these vegetation types, uninfluenced by fire, logging or other disturbances, are shown in Figure 11-5.

Subhabitats (level 5) within the major communities reflect spatial or species differences due to variances in hydroperiod, elevation, adjacent biota, fire, logging, etc. which strongly influence species distributions and rates of processes. For example, the aquatic bed and emergent persistent prairie communities contain six subsystems (Figure 11-4). The raw peat island, shrub island and tree island subhabitats represent seral stages observable in most aquatic bed prairies. Lakes are relatively autonomous subsystems. Shrub communities may be *Lyonia*-dominated, *Cyrilla*-dominated, or mixed. In mixed cypress stands, pond cypress may be associated with bays, blackgum or pine. Specialized subhabitats may also occur. The distinctive Mack's Island bird rookery behind the Suwannee River Sill is a variant of blackgum forest produced by guano inputs. Subhabitat components (level 6) are subsets of each subhabitat, usually defined by specific research. Four examples are given in Figure 11-4 for aquatic bed prairies: algae-*Utricularia* clumps (Bosserman,

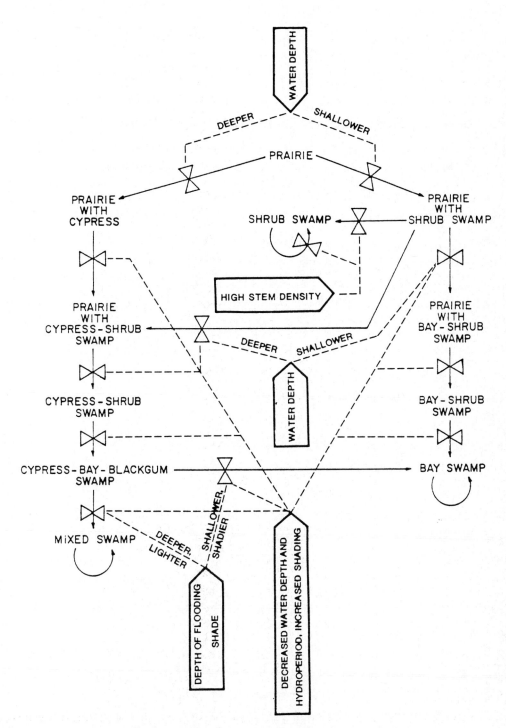

Figure 11-5. Autogenic succession in Okfenokee Swamp (Hamilton, 1982).

1979), alligator holes, peat and open prairies. Bosserman decomposed his clumps into several additional levels. So did Flebbe (1982) in the model discussed below.

The preceding phase 1 model and its precursors have served, and continue to serve, in several ways to loosely integrate stages 1 and 2 of the study plan. First, the levels in the model directly provide focal points for research. Second, interrelationships across levels and composition of higher level properties from lower level ones are brought into focus. Third, model based research leads to refinement and revision of the model itself. Finally, any research undertaken, no matter how specialized, has the overview model for context, and results of diverse investigations are readily assimilated into a developing integrated base of knowledge about the Okefenokee ecosystem.

Carbon Biogeochemistry Model

The water budget model (Patten and Matis, 1982) mentioned in the introduction is a phase 2 model (Figure 11-3) at level 1, the watershed ecosystem (Figure 11-4). Flebbe (1982) produced another phase 2 model to compare carbon dynamics in three Okefenokee major communities (level 4). This model's compartments were resolved to one level finer than subhabitat components (level 6).

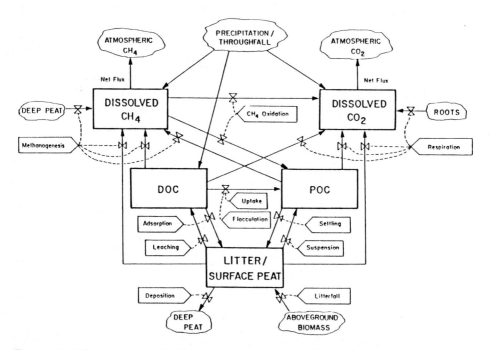

Figure 11-6. Conceptual model for carbon biogeochemical cycle in the aquatic subsystem of Okefenokee Swamp (from Flebbe, 1982).

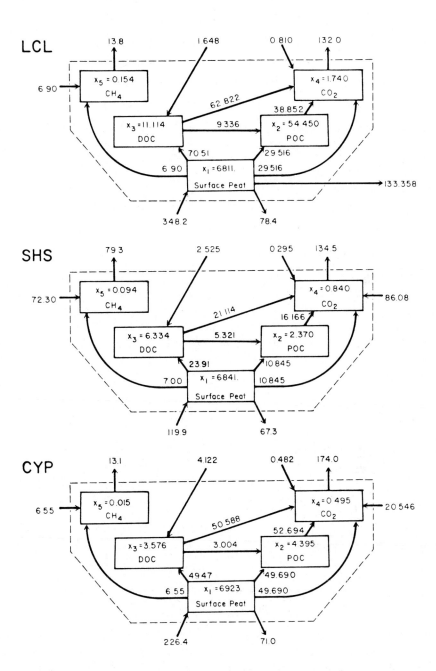

Figure 11-7. Quantified carbon models for the aquatic subsystem of three Okefenokee Swamp sites: LCL—Little Cooter Lake prairie; SHS—Chesser Prairie shrub swamp; and CYP—Chase Prairie cypress forest. Numbers within the five compartmental boxes represent standing stocks, or steady state storages (g C/m^2). Arrows are labelled with flows (g C m^{-2} y^{-1}) (from Flebbe, 1982).

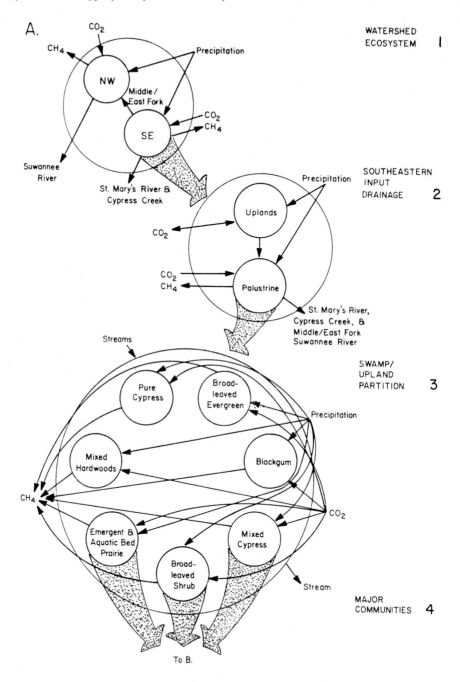

Figure 11-8. Aquatic subsystem carbon model embedded in the Okefenokee Swamp hierarchical model (Figure 11-4). Each diagram depicts successive links in the hierarchical model. The system level is noted above each diagram and the subsystem level below the diagram. Simple arrows denote flows of carbon within a given system, and large shaded arrows connect subsystems to systems at successive levels (from Flebbe, 1982). (*Continued on next two pages.*)

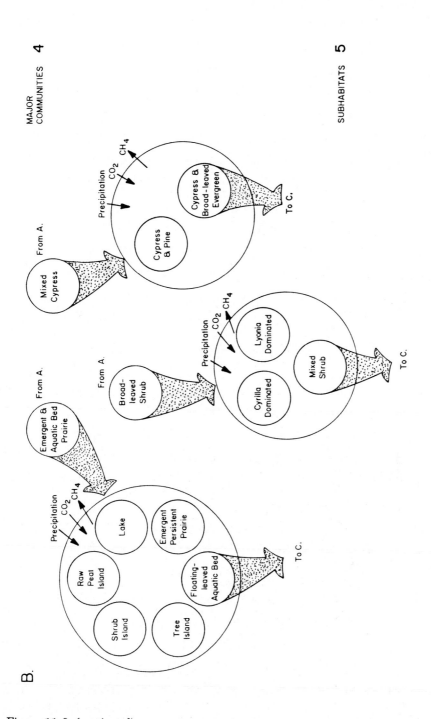

Figure 11-8. *(continued)*

C.

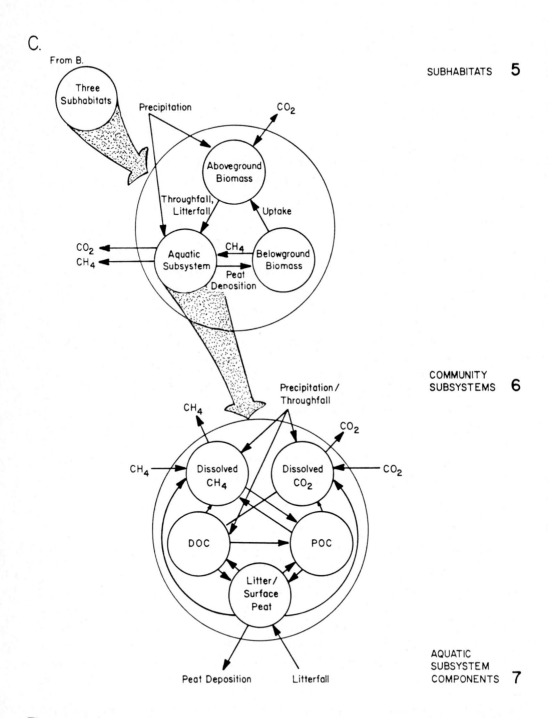

Figure 11-8. *(continued)*

That is, Flebbe defined three subhabitat components to use in modelling carbon biogeochemistry: aboveground portions of plants and associated biota, the below-ground peat subsystem including plant roots and procaryotes, and the aquatic subsystem. The last, the one of interest, was decomposed into five compartments whose carbon storages were each represented by a state variable x_i, i=1,....,5, taking values from a set X. Inputs z_i and outputs y_i to and from the level 6 environment receive values from two value sets, Z and Y, respectively. The model was thus constructed in the state space mode; Flebbe (1982) discussed the state transition and response functions, but they are not of concern here.

The model is shown conceptually in Figure 11-6, and quantified in Figure 11-7 for three sites representing Okefenokee major communities: Little Cooter Lake marsh (LCL), an aquatic bed prairie in Grand Prairie; a shrub swamp (SHS) in Chesser Prairie; and a mixed cypress forest (CYP) near Chase Prairie. The linkage of the Figure 11-5 model to the hierarchical model of the preceding section is detailed in Figure 11-8. The carbon storage and flow characteristics of the three communities in relation to one another are as follows:

Storages. The mass of peat and litter (x_1) is the same at each site. DOC (x_3), CO_2 (x_4) and CH_4 (x_5) all decrease from marsh (LCL) to shrub (SHS) to forest (CYP). POC (x_2) is 11-fold higher in the marsh than in the other two sites.

Inputs. Litterfall (into x_1) is greatest in the marsh and similar in the other two sites. Precipitation input to DOC (x_3) increases from marsh to shrub to forest. CO_2 (x_4) input through precipitation is several times higher for the marsh than for the shrub and forest sites. CH_4 (x_5) input from methanogenesis is 11-fold higher in shrub than in marsh or forest.

Outputs. The marsh differs in having CO_2 and CH_4 losses to the atmosphere through macrophytes, and no CO_2 input from roots. Deposition (from x_1) is similar at all three sites. CO_2 evolution (from x_4) is comparable, but slightly higher for the forest. CH_4 evolution (from x_5) is 6-fold higher for shrubs than for the other two sites.

Internal flows. No noteworthy differences exist in internal flows between the three sites.

Analysis of this model into environs is shown for selected cases in Figures 11-9 through 11-11. Flebbe (1982) should be consulted for other environ diagrams and more detailed treatment of results. Figure 11-9 illustrates the output environ generated by one unit of carbon input to the litter and peat compartment (x_1), and Figs. 11-10 and 11-11 show input environs that produce one unit of CO_2 and CH_4 outputs, respectively. In these figures, carbon flows are associated with arrows, and storages are the upper numbers in each box. Compartmental contacts (the number of times a unit of carbon enters a compartment while in the system) are given as means and, parenthetically, standard deviations by the middle numbers in each box. The lower numbers represent mean residence times (future for output environs, past for input environs), with parenthetical coefficients of variation. Statistical information is for the exponential distribution. Some points of comparison from the three sets of environs shown are as follows.

Output environ. Litterfall (Figure 11-9) leads to more carbon deposition to peat in the marsh (0.5613) than in the other two communities. It produces more CO_2 evolution form the forest (0.6575) and more CH_4 evolution from shrubs (0.0584). Peat conversion to POC and CO_2 is 3-fold higher in the forest (0.2195). Litterfall contribution to peat storage is least in the marsh (19.5606), to POC 11-fold higher there (0.1555), to DOC lowest in the forest, to CO_2 2-fold greater in the marsh, and to CH_4 11-fold higher in the marsh (0.0002). The future residence time of litterfall carbon in the peat compartment is highest for shrub (57.06 years) and lowest for marsh. In DOC it is highest for shrub (17.44 days) and lowest for forest. In CO_2 it is highest for marsh (1.79 days), and in CH_4 it is much lower for the marsh (1.94 hours vs. 36.37 months and 17.41 months for shrub and forest, respectively). In general, extremes appear in the marsh with greater frequency. The rapid turnover of marsh CH_4 generated by litterfall compared to that in the shrub and forest communities appears particularly notable.

Input environs. For CO_2 evolution (Figure 11-10), the carbon sources vary somewhat in the three habitats. Litterfall is the major source in marsh (0.9814), and to a lesser extent in forest (0.8555), but in the shrub community root uptake of dissolved CO_2 (0.6400) is the major source, with litterfall (0.3390) secondary. This reflects dominance in the shrub community by dense evergreen shrubs which defoliate slowly and in which dead twigs take a long time to fall through stems to the peat surface. The conversion of litter (x_1) to POC (x_2) and CO_2 (x_4) (0.0806, 0.1202) occurs more slowly than in the other two communities (note the long future residence times for peat and litter (x_1), 57.06 years, in the input environ for SHS in Figure 11-9). The conversion of litter (x_1) to DOC (x_3) (0.5342), and thence to POC (x_3) (0.0707) and CO_2 (x_4) (0.4759) is most rapid in the marsh community. In all three systems, little of the CO_2 evolved originates in precipitation. Also, all three communities are comparable in the amounts of storage, and past residence times of this storage, committed to ultimate CO_2 evolution. Concerning CH_4 evolution (Figure 11-11), most CH_4 (x_5) evolved by the shrub community is derived from deep (0.9117) rather than shallow (0.0883) peat, whereas the ratio is 50-50 for the other two systems. Unlike CO_2, no CH_4 originates as carbon in precipitation. The conversion of litter and shallow peat (x_1) to dissolved CH_4 (x_5) is much slower in shrub (0.0883) than in marsh or forest. The litter storage and past residence time per unit of CH_4 output are smallest in the shrub system and three times greater in the forest.

The comparative information depicted in Figures 11-9 through 11-11 illustrates the utility of environ analysis to extend an original data set to new relationships. The analysis methodology is only the expression of a broader theory, however, and it is the latter (Patten et al., 1976; Patten, 1978, 1982; Patten and Auble, 1981) which gives the Okefenokee program its unity through the dynamical system model carried forward to all possible research topics by the hierarchical model. This is an approach that could be recommended for much of ecosystem ecology.

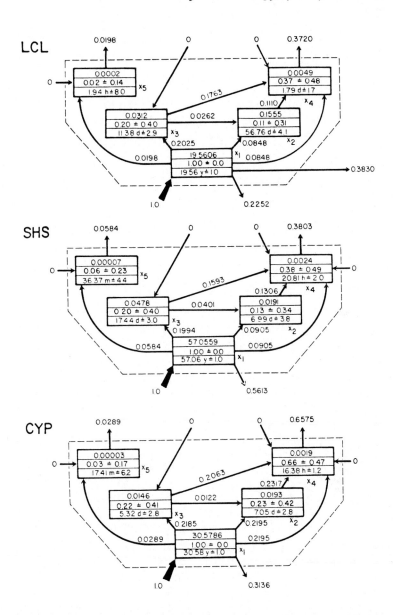

Figure 11-9. Unit output environs generated for the prairie (LCL), shrub (SHS) and cypress (CYP) sites by one unit of input (bold arrow) as litterfall to surface peat and litter (x1). Arrows are labelled with carbon flows (gC m^{-2} y^{-1}), and storages (g C/m^2) appear at the top of boxes corresponding to compartments in Figure 11-6. Middle numbers in the boxes are means plus or minus standard deviations (unitless) of the expected number of times carbon entering the system as litterfall will cycle through the compartment before exiting the system. Bottom numbers are means plus or minus coefficients of variation of the future residence times in the compartment of carbon entering the system as litterfall. Units for the latter means are y—years, d—days, h—hours, and m—minutes; coefficients of variation are unitless (from Flebbe, 1982).

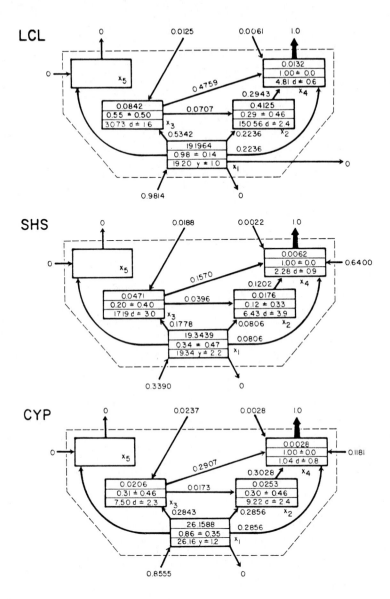

Figure 11-10. Unit input environs generating for the prairie (LCL), shrub (SHS) and cypress (CYP) sites one unit of output (bold arrow) from the carbon dioxide compartment (x_4). Arrows are labelled with carbon flows (gC m^{-2} y^{-1}), and storages (g C/m^2) appear at the top of boxes corresponding to compartments in Figure 11-6. Middle numbers in the boxes are means plus or minus standard deviations (unitless) of the expected number of times carbon exiting the system as CO_2 cycled through the compartment since entering the system. Bottom numbers are means plus or minus coefficients of past residence times in the compartment of carbon exiting the system as CO_2. Units for the latter means are y— years, d—days, h—hours, and m—minutes; coefficients of variation are unitless (from Flebbe, 1982).

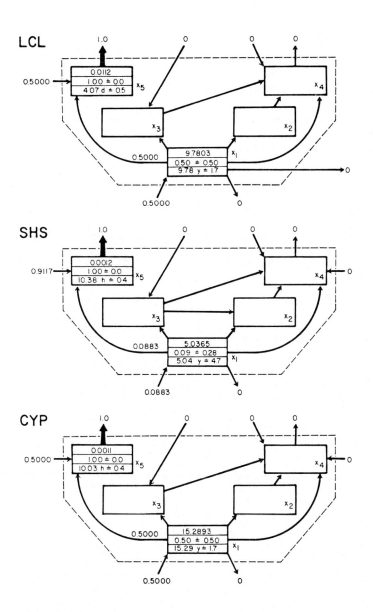

Figure 11-11. Unit input environs generating for the prairie (LCL), shrub (SHS) and cypress (CYP) sites one unit of output (bold arrow) from the methane compartment (x_5). Format and units are in Figure 11-10 (from Flebbe, 1982).

Systems Revision of Ecology

Ecology above the organism level has a compelling need to become a systems science. Most conceptual problems in theoretical ecology and all applied environmental problems are systems problems. They require the language and methods of systems for clear exposition and solution. When Clements (1916) wrote early in the history of plant ecology, "As an organism, the formation arises, grows, matures, and dies.... Furthermore, each climax formation is able to reproduce itself, repeating with essential fidelity the stages of its development," he was not conjuring a literal superorganism, he was using the most apparent biological metaphor for system available in the neo-Darwinian era. Gleason's (1926) reply was antisystem, an important if not dominant advocacy in much of ecology still today: "...an association is not an organism, scarcely even a vegetational unit, but merely a coincidence...merely the resultant of two factors, the fluctuating and fortuitous immigration of plants and an equally fluctuating and variable environment." Any distinction between structure, the species list, and integrated function, the communal organization that might have connoted "organism," went unrecognized by Gleason at the same time Elton (1927) was writing about constants in community organization: "The importance of studying niches is partly that it enables us to see how very different animal communities may resemble each other in the essentials of organization....There is often an extraordinarily close parallelism between niches in widely separated communities."

The niche has been extended to systems ecology by environs (Patten and Auble, 1981) and, as already mentioned, concepts of environment (Patten, 1978) and succession (Hamilton, 1982) have also been formulated in dynamical system terms. It should be possible to rewrite most of the book of ecology in this way to provide formal foundations to present empirical knowledge. Then, true laws of ecology will be possible. A state space treatment of Clement's climax can serve to illustrate the difference between an ad hoc ecology and a science of laws.

There are no climax plant communities in Okefenokee Swamp. Therefore, in this local context the climax concept is of questionable utility. Beyond Okefenokee, phytosociologists generally are beginning to doubt that climaxes exist anywhere on earth. So certain was Egler (1982) of their absence that he was willing to wager up to $10,000 against any believer in "plant succession to climax" who thought he could produce one. Egler will never have to pay because in nonformal vegetation science nothing can ever be pinned down and the rules of the game are subject to whimsical change. Avenues of escape are everywhere. In formal science this cannot happen.

On the issue of the existence of climax, systems ecology would ask first, what is the scale? If the Carboniferous forests developed by the same dynamic principles known today and reached climax, then where are they today? Clearly, succession to be meaningful must be considered on a shorter time scale than geological, where things always change—a major loophole for Egler. So, systems ecology provides a time frame T spanning relevant subintervals. What about spatial scales? Tansley (1935) found it necessary to introduce the polyclimax concept to account for vegetational mosaics in a region, each type equilibrated to local variations in environmental conditions. How local?— more nuance, more room for evasiveness. The systems approach responds by selecting a point and making a model there. Yes, it's an abstraction.

So, at a point there will exist a vegetation system modelled by Equation 11-1. The point can be expanded to different neighborhoods as and if required. What will the system do? Under influence of its inputs it will undergo determinate successional behavior to produce a sere on some interval of T, say [t,t"], where $t \leq t' \leq t"$. This means, as for all dynamical systems, that beginning with some initial state the system will pass through a sequence of transient (sooner or later vanishing) states and ultimately settle into a sequence of repeating (nonvanishing) steady states. If the initial state is a steady state, transient states will not occur in the sequence. Sometimes the number of steady states will only be one. Thus, what Clements discovered in the unlikely stretches of the North American prairie was not the ontogeny of a superorganism, but an already well known principle of general systems theory, one that had been formalized in mathematics a few centuries before in differential equations. Being well known and universal, ecologists should be utilizing such principles, not rediscovering them never knowing they have done so, and never knowing their linkages to others.

Now, consider the sere on $t \leq t' \leq t"$. With u = x,y,z, where x, y and z are state, output and input variables which take values, respectively, from X,Y and Z in 11-1, let u(t') represent the value of u at t', $u^{tt'}$ be the history of values from t to t', and $u_{t't"}$ be the future of values after t' through t". If the vegetation state x(t') at t' is altered to another state **x**(t'), which is equivalent to an earlier state, say x(t), then the system has regressed to an earlier successional stage, a well known disturbance phenomenon in plant ecology, with recovery dynamics to follow the discontinuous change of state. Why discontinuous? Because the state x(t') of a fixed structure system (Astor et al., 1976, p. 391) was transformed instantly to **x**(t'), a direct state or initial conditions type of perturbation (e.g., Astor et al., 1976, p. 401). Alternatively, a disturbance $z_{(t'-\Delta t)t'}$ beginning some time Δt before t', lasting to t', and leaving the system in x(t'), could have effected the change as an input or general parameter type perturbation (Astor et al., 1976, p. 391). The state change that results is continuous in this case. The last fires in Okefenokee until 1981 were during 1954-55. In 1981 several small fires occurred. On a time scale of 30 years, vegetation at the affected points went from a pre-burned to a post-burned state in a few minutes. Whether this is recorded as a discontinuous or continuous change depends on the time resolution of observation only. What happens next?

If the two intervals [t,t'] and [t',t"] are of the same length, and if the input sequence $z_{t'(t"-\Delta t)}$ is equivalent to what it was earlier on [t,t'-Δt], i.e., $z_{t'(t"-\Delta t)} = z_{t(t'-\Delta t)}$, then succession from t' to t"-Δt will repeat what it was during t to t'-Δt.

The postdisturbance state $x(t''-\Delta t)$ will be equivalent to the predisturbance state $x(t'-\Delta t)$. The system will have recovered. Almost certainly, however, $z_{t(t'-\Delta t)} \neq z_{t'(t''-\Delta t)}$. To greater or lesser degree there will generally be environmental differences between the two periods, leaving the final states unequal, $x(t'-\Delta t) \neq x(t''-\Delta t)$, at some level of discrimination. If the difference between $z_{t(t'-\Delta t)}$ and $z_{t'(t''-\Delta t)}$ is small, then the difference between $x(t'-\Delta t)$ and $x(t''-\Delta t)$ would also be small, and whether or not these two states would be considered equivalent depends again upon resolution. Thus, resolution is another major loophole working for Egler. If $x(t'-\Delta t) \neq x(t''-\Delta t)$, the vegetation would be judged not to have recovered at $t''-\Delta t$ from the perturbation during $[t'-\Delta t, t']$.

This should not obscure the fact, however, that the propensity to recover is lawfully inherent given that $x(t') = x(t)$. Furthermore, cycles of perturbation and recovery are possible so long as the perturbations are to state or input, respectively. Even so, the real situation will not be so simple as set forth above. Real disturbances occur to any states and at any times, and observation intervals and resolution (sampling and analysis) methods vary. Thus, the measured vegetation at any point on a landscape represents the historical integration of all these factors, carried forward lawfully to the moment of observation, and depends also on the tools of observation. The adjacent point may differ in particulars, and so on, resulting in the commonly observed mosaic distribution at macroscales.

Transition function perturbations of variable structure type (Astor et al., 1976, p. 391) are more basic. These involve, for example, genetic mutations and, from these, selection of a new community genome. While such phenomena are at work continuously, long time scales are generally required before they can be resolved as successionally relevant. As observed earlier, evolutionary time is much longer than successional time, and normally the two processes are not perceived to intersect. If they should, however, then some modification of Equation 11-1 to account for transition function changes during successional time T would have to be introduced.

Now, what about climax? Let $t \leq t' \leq t''$ span an entire sere from initial stage $x(t)$ to climax $x(t')$, and then beyond. By Clementsian dicta, a sampling of the state of this community at time t'' should yield $x(t'') = x(t')$. The climax, once achieved, is supposed to be invariant. Suppose $x(t'') \neq x(t')$. This could only mean one of two things in dynamical system theory: (1) an evolutionary change occurred during $[t', t'']$, changing ϕ, or (2) the input sequence $z_{t't''}$ is not equivalent to that which produced the climax $x(t')$, i.e., the environment changed. Evolutionary change is normally expressed in geological time, not successional time. Thus, in general, an environmental change has occurred on $[t', t'']$. This change may be of two types, steady state or otherwise. Steady state change could correspond, for example, to climatic cycles which would be reflected in steady state vegetation cycles. The vegetation at a point could change, but each type would represent a stage in the harmonic steady state. It is questionable whether vegetation undergoing this kind of dynamics would be considered, or could be recognized over the long time scales involved, as climax, although it would certainly be steady state. Here is another major loophole for Egler, who can simply argue that f it changes it's not climax. Nonsteady state change in environment, again difficult to distinguish from the first case for the long time scales involved, would initiate a nonsteady state sequence of vegetation dynamics. Climax vegetation would then have been replaced by successional vegetation.

If above, $x(t')$ were a disturbance disclimax, such as a fire-maintained shrub, bay, or cypress-bay-blackgum swamp in Okefenokee (Hamilton 1982), then removal of the disturbance regime during [t', t''] would release the system for progression to a normal, nondisturbance climax which might be attained by t'', i.e., $x(t'')$. Then, $x(t'') \neq x(t')$ represents the endpoint of a response which is mandated, but would be described by many ecologists as "adaptive" to the new conditions. Adaptation theory is another overworked area of ecology that a general systems approach would also significantly reform, but that is material for another paper. The issue here is, does the climax exist? If there are no examples of climax vegetation on earth, what should become of climax in the conceptual repertoire of ecology? Most ecologists, and this is only a guess, would have it expunged to join the ranks of ether, phlogiston, homunculi, gemmules and other myths of earlier science. Without a physical realization, of what possible use is a concept in a strictly empirical field? Systems ecology, a formal science, would argue that climax is a concept applied at a certain hierarchical level of the organization of matter on earth to represent a set of phenomena universal to all levels of organization. Even if the ideal is never physically realized, its preservation as a concept linking vegetation dynamics to all system dynamics does more to explain the mantle of plant cover that populates the globe than all the descriptions of vegetation anywhere could possibly ever do.

In ad hoc, empirical science there is no need for ideal gases and frictionless pendulums or pulleys, but in formal science that seeks to discover laws of nature behind empirical reality, there is. The Okefenokee study has as premises the following: (1) that it is improvident for ecology to attempt to advance solely through continued rediscovery of old principles in new realms, or worse, to miss the principles entirely and record only the details of their expression; and (2) that the reformulation and consolidation of ecology according to already known laws of systems are necessary antecedents to the discovery of genuinely new laws of ecology. Accordingly, the research program in Okefenokee is geared to providing as formal an ecology of the great swamp as the current state of ecosystem science, activated by inputs from general systems theory, will allow.

Acknowledgements

This is University of Georgia Contributions in Systems Ecology, No. 71 and Okefenokee Ecosystem Investigations, Paper No. 67, supported by grants from the U.S. National Science Foundation.

References

Astor, P.H., B.C. Patten, and G.N. Estberg. 1976. The sensitivity substructure of ecosystems. Pages 388-429 in Patten, B.C., editor. *Systems Analysis and Simulation in Ecology, Vol. 4*. Academic Press, New York.

Auble, G.T. 1982. *Biogeochemistry of Okefenokee Swamp: litterfall, litter decomposition, and surface water dissolved cation concentrations*. Ph.D. dissertation, University of Georgia, Athen, 311 pages.

Barber, M.C. 1983. *Nutrient dynamics of orb weaving spiders (Araneae, Araneidae) in Okefenokee shrub swamps*. Ph.D. dissertation, University of Georgia, Athens, 192 pages.

Blood, E.R. 1981. *Surface water hydrology and biogeochemistry of the Okefenokee Swamp Watershed*. Ph.D. dissertation, University of Georgia, Athens, 194 pages.

Bosserman, R.W. 1979. *The hierarchical integrity of Utricularia - Periphyton microecosystems*. Ph.D. dissertation, University of Georgia, Athens, 266 pages.

Clements, F.C. 1916. *Plant Succession: An Analysis of the Development of Vegetation*. Carnegie Inst. Washington Publication 242:1-512.

Egler, F.E. 1982. Letter to the editor. Bull. Ecol. Soc. Amer. 62:230-232.

Elton, C. 1927. *Animal Ecology*. Sidgwick and Jackson, London.

Flebbe, P.A. 1982. *Biogeochemistry of carbon, nitrogen and phosphorus in the aquatic subsystem of selected Okefenokee Swamp sites*. Ph.D. dissertation, University of Georgia, Athens, 348 pages.

Gill, A. 1962. *Introduction to the Theory of Finite State Machines*. McGraw Hill, New York.

Glasser, J.E. 1983. *Pattern, diversity and succession of vegetation in Chase Prarie, Okefenokee Swamp: A hierarchical study*. Ph.D. dissertation, University of Georgia, Athen, 201 pages.

Glasser, J.E. and M.C. Barber. 1983. Formulating hydroperiod effects for a multispecies stand simulation model for the Okefenokee Swamp. Pages 733-769 in W.K. Lauenroth, G.V. Skogerboe and M. Flug, editors. *Analysis of Ecological Systems: State-of-the-Art in Ecological Modelling*. Elsevier, Amsterdam.

Gleason, H.A. 1926. The individualistic concept of the plant association. *Bull. Torrey Bot. Club* 53:7-26.

Hamilton, D.B. 1982. *Plant Succession and the influence of disturbance in the Okefenokee Swamp, Georgia*. Ph.D. dissertation, University of Georgia, Athens, 254 pages.

Klir, G.J. 1969. *An Approach to General Systems Theory*. Van Nostrand Reinhold, New York.

Matis, J.H. and B.C. Patten. 1981. Environ analysis of linear compartmental systems: the static, time invariant case. *Bull. Int. Stat. Inst.* 48:527-565.

Mesarovic, M.D., D. Macko and Y. Takahara. 1970. *Theory of hierarchical, multilevel systems*. Academic Press, New York.

Mesarovic, M.D. and Y. Takahara. 1975. *General System Theory: Mathematical Foundations.* Academic Press, New York.

Monk, C.D. 1968. Successional and environmental relationships of the forest vegetation of north central Florida. *Am. Midl. Nat.* 79:441-457.

Pattee, H.H., editor. 1973. *Hierarchy Theory.* George Braziller, New York.

Patten, B.C. 1978. Systems approach to the concept of environment. *Ohio J. Sci.* 78:206-222.

Patten, B.C. 1982. Environs: relativistic elementary particles for ecology. *Am. Nat.* 119:179-219.

Patten, B.C. and G.T. Auble. 1981. System theory of the ecological niche. *Am. Nat.* 118:345-369.

Patten, B.C., R.W. Bosserman, J.T. Finn and W.G. Cale. 1976. Propagation of cause in ecosystems. Pages 457-579 in B.C. Patten, editor. *Systems Analysis and Simulation in Ecology, Vol. 4.* Academic Press, New York.

Patten, B.C. and J.H. Matis. 1982. The water environs of Okefenokee Swamp: an application of static linear environ analysis. *Ecol. Modelling* 15:1-50.

Rykiel, E.J. 1977. *The Okefenokee Swamp watershed: water balance and nutrient budgets.* Ph.D. dissertation, University of Georgia, Athens, 246 pages.

Stinner, D.H. 1983. *Colonial wading birds and nutrient cycling in the Okefenokee Swamp.* Ph.D. dissertation, University of Georgia, Athens. 129 pages.

Tansley, A.G. 1935. The use and abuse of vegetational terms and concepts. *Ecology* 16:284-307.

Windeknecht, T.G. 1971. *General Dynamic Processes, a Mathematical Introduction.* Academic Press, New York.

Wymore, A.W. 1967. *A Mathematical Theory of Systems Engineering: The Elements.* Wiley, New York.

Zadeh, L.A. and C.A. Desoer. 1963. *Linear System Theory. The State Space Approach.* McGraw-Hill, New York.

Ziegler, B.P. 1978. Structuring the organization of partial models. *Int. J. Gen. Syst.* 4:81-88.

12/ SUMMARY AND STATE OF THE ART OF WETLAND MODELLING

William J. Mitsch
Milan Straškraba
Sven E. Jørgensen

Characteristics of Wetland Models

The various approaches to the modelling of wetlands, as described in this book, reflect, to some degree, approaches taken for other aquatic and terrestrial systems. Furthermore, the development of wetland modelling is still in its infancy, with new approaches combining with the old. But there is not a complete transfer of earlier modelling techniques for wetlands, nor is the development of wetland models limited by prior paradigms. This is due to some of the unique features of wetlands and to a variety of issues that are involved in wetland management. Described below are some of the important distinctions and features of wetlands that make them different systems to model.

1) **Diversity of Wetland Types**. Wetlands are frequently viewed as ecotones—transitional zones between aquatic and terrestrial ecosystems—and they differ from both types of systems. There is also a diversity and heterogeneity of wetland types, with a greater range of structural and functional features than those found with other ecosystems. Wetland systems range from low productivity, peat-building bogs to highly productive, carbon-exporting salt marshes, from monospecific reedswamps to diverse bottomland hardwood wetlands. Furthermore, because of human intervention, created or altered wetlands are now a frequent part of the landscape. Each of these wetland types represents a unique system for modelling, even with common ecological principles uniting them. The diversity of types of wetlands has often resulted in wetland models being limited to specific regions or types of wetlands. As an example, the forested wetland models in Chapter 7 are significantly different than the salt marsh models discussed in Chapter 5 in terms of model structure and rate

constants. Both of these groups of models contrast with mathematical models developed for lakes and reservoirs (described in Chapters 9 and 10) which are homogeneous and dilute systems relative to wetlands. These lake and reservoir models are more easily adapted to systems ranging over wide geographic and trophic status (see, e.g., Straškraba and Gnauck, 1985).

2) **Hydrochemical Complexity of Wetlands**. The physical factors controlling wetlands, particularly hydrology, are very important and complex. The chemical and biological characteristics of wetlands are closely tied to the hydrologic conditions, including many features of the hydroperiod such as flooding frequency, water depth, and seasonal patterns of flooding. This physical complexity is also manifest in the complexity of the hydrologic budget of a wetland, as described in Chapter 2. Some components of the hydrologic budget, particularly the groundwater flows and evapotranspiration, require much more study in the field and laboratory to quantify these relationships for wetland models. Because hydrology is so important to the structure and function of wetlands, wetland models require fairly accurate descriptions of the hydrologic budget. The models must also be able to describe the effects of hydrology on chemical and biological processes. For example, it is important to understand the interactions among hydrology, nutrient conditions, and wetland productivity when describing wetlands as nutrient retention systems (see Chapter 8).

3) **Time Constants and Transitional Wetlands**. Wetlands are constantly changing, due to both changing external hydrologic conditions and gradual changes in organic storages (e.g. hydrarch succession). When the flooding duration of a forested wetland is increased, it may change to a herbaceous or shrub wetland. Further inundation could cause it to become an algal-dominated open water system. Each of these systems is vastly different in the structural characteristics and, most importantly for modelling, in the dominant time constants for vegetation. The extent of mosaics of water, grasses, shrubs, and trees change constantly as seen in systems such as the Okefenokee Swamp (Chapter 11).

4) **Interfaces**. Interfaces play an important role in wetland dynamics but they are generally more complex and less understood. Typical examples are sediment-water exchange of chemical species; interactions of water and soil, soil water and plants, and free water solute and submerged vegetation; and the air-water interface (including micrometeorological effects). Wetlands are by definition shallow or intermittently flooded systems. Because of the proximity of the water column to the sediments, one medium greatly affects the other and exchanges of chemicals are complex phenomena. Processes such as sorption, cation exchange, mineralization, sedimentation, resuspension, denitrification, and leaching can all be important in certain wetlands are are all related to this water-sediment interface. Much needs to be translated from existing

literature on shallow lake ecosystems models to suitable algorithms for other wetlands. There has been more success in dealing with sediment-water exchange phenomenon in shallow water bodies (see, e.g., Chapter 10), although this remains one of them most difficult aspects of shallow water bodies to model with consistent accuracy.

5) **Exchanges with Adjacent Ecosystems**. The interface question is also significant for wetlands because of the dependence of wetlands on outputs from neighboring ecosystems and interrelations with them. Often the boundaries of wetlands are ill-defined or hydrologically complex. In some instances, the boundary between upland and wetland is not fixed, as the wetlands increase or decrease in size with changing hydrologic conditions. These phenomena point to the importance of wetland modelling on a spatial scale, as exemplified by the spatial coastal model discussed in Chapter 6.

State of the Art of Wetland Modelling

Wetland modelling has advanced significantly since one of the authors described the state of the art of freshwater wetland modelling in the early 1980s (see Mitsch et al., 1982, Mitsch, 1983) and a subsequent review of freshwater wetland model articulation, accuracy and effectiveness was published by Costanza and Sklar (1985). Modelling efforts for wetlands were also discussed for the seven major classes of wetlands in a recent wetlands textbook (Mitsch and Gosselink, 1986), but, because the broad scope of the textbook included ecological details of all types of wetlands, little detail of specific models was given. The present book, which gives a much better demonstration of the present state of the art of wetland modelling and includes coastal saltwater wetlands as well as freshwater types, greatly complements that textbook.

The present status of wetland modelling is summarized in Figure 12-1. Many advances are evident over the past half decade. The status of coastal wetland models, including most salt marsh and shallow estuary models, has been summarized in Chapters 5 and 6 and is well developed along several approaches. The use of spatial wetland models is a particularly noteworthy advancement, and one which may ultimately dominate wetland modelling, particularly with the advances in graphics capabilities on mainframe and micro computers. Why there have not been more systems studies of mangrove wetlands is a puzzle to us, since these wetlands dominate many coastal regions of the tropical and sub-tropical world.

Bog and northern peatland models as exemplified by the work in Chapters 3 and 4, when added to the abundant modelling efforts in North America and Europe in tundra ecosystems, make a significant body of modelling literature on peat-dominated cold-climate wetlands. Models of freshwater marshes (which includes a range of wetlands from wet meadows to reedswamps to *Typha* marshes) are a more diffuse group, and we now feel that this group of wetlands, along with tropical mangrove wetlands, may be the most under-represented wetlands in mathematical modelling efforts. Part of the reason is the great diversity of freshwater marsh wetlands, but other reasons include the relative lack of recognition of the importance of these systems relative to northern peatlands, forested wetlands, and coastal salt marsh wetlands. Chapter 8 gives the start to a generic nutrient balance model for these types of wetlands.

There will be more use of specialized model-building software for models of all types, including wetlands. The modelling effort in Chapter 7, using the software STELLA is an example of such a technique. Forested wetland modelling, as described in that chapter, has been modest and has developed in one major direction, emphasizing terrestrial tree growth models adapted to flooding conditions. Chapters 9 and 10 describe the modelling efforts of shallow lakes and reservoirs, a group of aquatic systems that collectively has a much richer history of modelling, primarily due to efforts to understand and control eutrophication. It is not surprising that shallow lakes and reservoirs have received so much attention, as the shallow bodies of water are the first to show symptoms of pollution problems such as eutrophication.

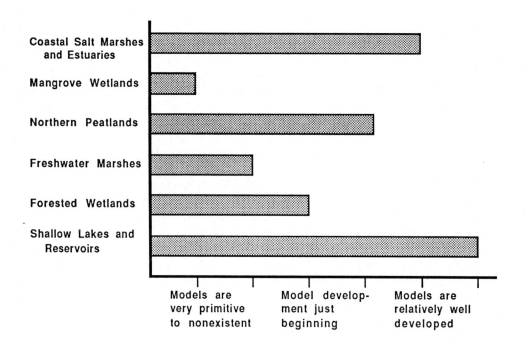

Figure 12-1. Relative effort of ecological modelling of various types of wetlands as of the mid 1980s.

Future Research Needs

Wetland modelling has now progressed to where significant new approaches are being applied to these systems and new questions keep emerging on these dynamic systems. But we need to know much more about these systems before we can adequately model them. What are the quantitative descriptions of processes in ecotones in general and wetlands in particular? What are the interrelations between different existing ecosystem models to allow coupling of different existing models into an overall wetland model? Is there a standard hydrologic model for wetlands? The commonly used models of subsurface flows in waterlogged soils and models for interrelations between chemistry and hydrology have to be evaluated in this regard. What are the effects of standing water and subsequent anaerobiosis on rates of chemical transformations and on productivity of wetland plants and other organisms? How does the wetting and drying cycle of wetlands affect physical, chemical and biological cycles? How important are internal biological control mechanisms relative to forcing functions? What is the importance of catastrophic events such as floods and hurricanes on wetlands and how do we model them? Are they deterministic or probabilistic? What about the effects on the wetland of altering the frequency of such events?

As with most modelling efforts in ecology, analysis of the uncertainty is an important aspect of wetland modelling. During recent years, more rigorous modelling procedures have been developed and recommended for ecosystem analysis (see, e.g., Jørgensen, 1986). However, when large-scale systems are modelled, a certain portion of "art" is still needed in addition to the science. Modelling may be approached from interpretation of data, from examination and knowledge of processes, and from theory of system functions. All three approaches should be applied to wetland modelling, although in practice a particular approach may be stressed. The modelling purpose will determine which combination of approaches is most appropriate.

In respect to the overall complexity, wetlands have to be considered systems of systems and appropriate methodology has to be developed and used. Holistic approaches, not seeking details but general rules, are much more important than for simpler systems where adequate details can be incorporated into models more easily. The interest for holistic approaches is, in general, increasing and it is expected that modelling approaches based on systems theory will be more widely used in the future. This will go along with further development of ecological theories, to which non-systems ecologists as well as ecological modellers have to contribute. Hierarchical models, methods of decomposition into interrelated subsystems and other complex methodologies will find here fruitful territory. The validity and applications of hierarchy theory, described in Chapter 11, provide particularly fertile ground for wetland modelling research in a theoretical framework. Can empirical and theoretical approaches be better combined? Can a vegetational mosaic, as is often the case in a regional wetland system, be treated as one system?

Most importantly to some, we need to further assess the importance of wetland systems in the landscape and how our actions affect these systems (conservationist's view) and how these systems affect human systems (ecotechnologist's view). This can be done through modelling with economic and ecotechnologic approaches that complement the mathematical approaches. Our models need to be simple as well as complex so that wetland managers can use them effectively to conserve this vanishing natural resource. We believe that, with proper modelling efforts focused on wetland management and theory, the benefits will be to humanity and wetlands alike.

References

Costanza, R. and F. H. Sklar. 1985. Articulation, accuracy, and effectiveness of mathematical models: a review of freshwater wetland applications. *Ecol. Modelling* 27:45-68.

Jørgensen, S.E. 1986. *Fundamentals of Ecological Modelling*. Elsevier, Amsterdam, 389 pages.

Mitsch, W.J. 1983. Ecological models for management of freshwater wetlands. Pages 283-310 in S.E. Jørgensen and W.J. Mitsch, editors. *Application of Ecological Modelling in Environmental Management, Part B.* Elsevier, Amsterdam.

Mitsch, W.J., J.W. Day, Jr., J.R. Taylor, and C. Madden. 1982. Models of North American freshwater wetlands. *Int. J. Ecol. and Environ. Sci.* 8:109-140.

Mitsch, W.J. and J.G. Gosselink. 1986. *Wetlands*. Van Nostrand Reinhold, New York, 539 pages.

Straškraba, M. and A.H. Gnauck. 1985. *Freshwater Ecosystems: Modelling and Simulation*. Elsevier, Amsterdam, 309 pages.

INDEX